GREAT AUK ISLANDS

A field biologist in the Arctic

For Francesca, Laurie and Nick and my parents

David Quinn

GREAT AUK ISLANDS

A field biologist in the Arctic

by Tim Birkhead

Illustrated by
DAVID QUINN

T & A D POYSER

London

First published 1993 by T & AD Poyser Ltd
Print-on-demand and digital editions published 2010 by T & AD Poyser, an imprint of A&C Black Publishers Ltd, 36 Soho Square, London W1D 3QY

www.acblack.com

ISBN (print) 978-1-4081-3786-4
ISBN (epub) 978-1-4081-3785-7
ISBN (e-pdf) 978-1-4081-3784-0

A CIP catalogue record for this book is available from the British Library

Typeset by Phoenix Photosetting, Kent

Printed in Great Britain by Martins the Printers, Berwick upon Tweed

Contents

Acknowledgements

This book had its inception a long way from the Arctic amidst phainopeplas, tarantula hawks and diamondbacks in the Usery Mountains in the Sonoran Desert, Arizona. I thank John Alcock for showing me the desert and for unwittingly providing the final impetus for me to start writing.

The direct and indirect help one receives in writing a book of this sort, which spans many years, is enormous and involves a large number of people. My first debt is to David Nettleship, with whom all my research in the arctic regions was conducted. His direct and indirect contributions to this book are immeasurable. I am extremely grateful to David for the financial support he obtained which allowed me to work at various seabird colonies throughout the eastern Canadian Arctic. I also thank him for sharing his extensive knowledge of arctic history, providing references and photographs, for allowing me to use his superb personal library, and, with his wife Angela and daughter Trisha, for generous hospitality.

This book is one of several alternative accounts that could have been written and I am sure the people who worked with me in different parts of the Arctic will have a rather different perspective. I also know that had an anthropologist lived among us in our various field camps his or her account would have been different again. The contribution made by those individuals that worked as field assistants was immense and I am grateful to them all but especially the late Gordon Calderwood, Bill Carson, Keith Clarkson, Ted Currie, Richard Elliot, Sid Elson, Anne Greene, Erick Greene, Sue Johnson, Ruth McLagen, Mike Malone, Randy Milton, the late Larry Patterson, Don Reid and Eric Verspoor. Erick Greene's contribution in particular was indispensable.

My colleagues and research students in the Department of Animal and Plant Sciences at the University of Sheffield were an irregular source of inspiration, and in this respect I would like to thank Len Hill, Kate Lessells, David Lewis, Mike Siva-Jothy and Phil Warren. From elsewhere I thank Anne Grebby and Paul Hodges for their subliminal contributions.

Others who have helped in different ways include Mike Alexander, Mick Ashby (Zoology Museum, Cambridge), John Biggins, Mike Bradstreet, Fred Breummer, David Cairns, John Croxall, Joyce Dagnall, Nick Davies, Bill and Millie Elson, Bobbie Fletcher, Tony Gaston, Dave Hollingworth (for photographic advice), David Macdonald, Richard Mearns, Pat Monaghan, Bill Montevecchi, Richard Morris, Christer Wiklund and Sherry Wilson. I am especially grateful to Jayne Pellatt for years of unstinting assistance and especially for drawing all the figures for this book.

Two good friends, Louise Hiom and Gordon Calderwood, lost their lives while studying seabirds. This book is dedicated to their memory.

One of the pleasures of doing research is the unexpected discovery of links between apparently unrelated bits of information. While I was researching some of the background information for this book in Cambridge I was fortunate enough to be able to view the fine collection of Great Auk eggs that Alfred Newton donated to the Zoology Museum in the late 1800s. Newton was a founder member of the British Ornithologists' Union and his library is now held within that of the Zoology Department. An hour or so after looking at the Great Auk eggs I was browsing through his collection of ancient scientific papers when I unexpectedly came across a signed reprint of Richard Owen's description of the Great Auk's skeleton, published in 1864 (see Chapter 4) which he had sent to Newton. The presence of that paper bearing Owen's signature formed a powerful link between past and present—the extinct and the extant.

I would like to thank all those in Cambridge, both in the Zoology Department and at the Scott Polar Institute Library, who allowed me to use their facilities. Linda Birch and Euan Dunn deserve special thanks for finding obscure material for me in the Alexander Library, Edward Grey Institute, Oxford. Other individuals and institutions which provided information in one form or another include the Public Record Office in London, Lieutenant A. C. F. David of the Hydrographic Department (Ministry of Defence), J. D. Brown of the Naval Historical Branch (Ministry of Defence), Mrs Shane Winser (Expedition Advisory Centre), Memorial University Library, St. John's Newfoundland, the Zoology Museum in the University of Manchester and the Moravian Historical Society: my thanks to them all.

During the past 20 or so summers I have visited and worked on a number of seabird islands, but none more often than Skomer Island, Wales. My final acknowledgments are to individuals who, in different ways, have helped to make Skomer a special place for me. I thank David Quinn for taking the time to visit Skomer to make some of the drawings for this book: as always his artwork is outstanding. In the distant past Andy Richford of T & A.D. Poyser worked on Skomer at the same time as myself: I thank him for his encouragement. I am grateful to Miriam Birkhead and Euan Dunn for being on Skomer at the right times, and more recently for reading through the entire manuscript and making numerous helpful comments. Finally, I thank my parents for their continued support and encouragement.

CHAPTER 1

Horizon Opening

Nothing in the whole system makes sense until the natural history of the constituent species becomes known. The study of every kind of organism matters, everywhere in the world.

Wilson (1987)

As the men I have named clambered up they saw two Gare-fowls [Great Auks] sitting among numberless other rock-birds (*Uria troile* and *Alca torda*), [Common Guillemots and Razorbills], and at once gave chase. The Gare-fowls showed not the slightest disposition to repel the invaders, but immediately ran along under the high cliff, their heads erect, their little wings somewhat extended. They uttered no cry of alarm, and moved, with their short steps, about as quickly as a man could walk. Jon with outstretched arms, drove one into a corner, where he soon had it fast. Sigurdr and Ketil pursued the second, and the former seized it close to the edge of the rock, here risen to a precipice some fathoms high, the water being directly below it. Ketil then returned to the sloping shelf whence the birds had started, and saw an egg lying on the larva slab, which he knew to be a Gare-fowl's. He took it up, but finding it was broken, he put it down again. Whether there was not another egg is uncertain. All this took place in much less time than it takes to tell it. They hurried down again, for the wind was rising. The birds were strangled and cast into the boat. . .

So reads Alfred Newton's (1861) account of that fateful day in early June 1844 on the island of Eldey, off southwest Iceland, when the last Great Auks were seen alive. The corpses of these two birds were sold to a dealer. The whereabouts of their skins is unknown, but (remarkably) some of their internal organs were preserved and are now in the Museum of Zoology at the University of Copenhagen, Denmark.

The Great Auk is a symbol: a symbol of man's greed, short-sightedness and, in the archaic sense of the word, awful ability to ignore apocalyptic warnings. The Great Auk is also a symbol of lost scientific opportunities. How much more we would have known about life, and about marine birds in particular, had the Great Auk still been extant.

How, you might ask, could the study of living Great Auks tell us more about life? The answer to that question forms part of this book, but we can also answer it in another way. For non-scientists the research done so enthusiastically by scientists must often appear desperately specialized, and it must sometimes also seem trivial. But trying to figure out what scientists do is rather like watching ants (Thomas 1974). If you sit down beside a large colony of wood ants and focus closely on just one individual you might see it dragging a pine needle, stumbling over twigs, and falling into minor crevasses. As far as you can see it appears to be moving around in an uncoordinated fashion. However, if you put the magnifying glass away and step back to look at the large and complex nest which the colony has constructed, then you begin to realize that all that seemingly random, individual effort is not for nothing. Although the contribution of each ant appears trivial, the combined effort of the entire colony produces something substantial, tangible and impressive. In just the same way many modest contributions add to the store of human knowledge, a process which 'has been the secret of Western science since the seventeenth century, for it achieves a corporate, collective power that is far greater than any one individual can exert' (Ziman 1969).

Few books are written about how scientists work. Scientists are often seen as a different breed from the rest of humanity and what they do often appears to be unintelligible to the lay-person. The public's image of scientists is made worse by the media who often portray them, at best as cold, calculating and perfectly objective, interested only in 'their' problem, and at worst as eccentric. Of course, some scientists are all of these, but a great many are not. Because scientists are forced to tell the world about their findings through the unexciting channels of the scientific literature, it is hardly surprising that their work often appears terse and dull. 'To the non-scientific outsider the prose style of the standard scientific paper gives the impression that the paper has in some curious way not been written by a human being' (Charlesworth *et al.* 1989). But there is no need for scientists to write in a boring or esoteric way. I have always felt that even relatively complicated biological ideas can be explained simply and I tell my students to write their PhD theses and research papers so that their parents can understand them (on the basis that most of them arise from non-scientifically trained stock). Unfortunately, even if scientists adopt this type of approach, the way in which they go about their business can still be difficult to appreciate, no doubt partly due to the shortage of space in scientific journals and the telegraphic style of writing we are forced to adopt.

Although many would not admit it, I believe that scientists often dislike their own style of writing. A habit I have noticed among my colleagues is that even if they do not read a scientific paper from beginning to end, almost all of them read the 'acknowledgements'. This is the short section at the end of a paper where the author (or authors) can thank the various people that have helped them in different ways with their study. Why are these acknowledgements so fascinating? I think the answer lies in the fact that this is often the only opportunity for scientists to try to inject a glimmer of humour or humanity into their writing. The acknowledgements section is thus the only 'window' the reader has through which to try and catch a glimpse of the author as a real person.

Science has been described as 5% inspiration and 95% percent perspiration. While this is largely true, it belies the effort involved in coming up with an original idea in the first place, but there is no doubt that testing ideas can entail a lot of hard and repetitive work. Long hours of field work, often carried out under dangerous, boring or exciting conditions, are usually summarized in a few dull sentences, in the name of objectivity. At an early stage in my career I read a paper in the journal *Animal Behaviour* on the social behaviour of orang-utans. In contrast to most other papers I had read, this one described in more than usual detail the potential problems of working in the Sumatran rainforest, including tangles with snakes and other predators. This brought the paper alive for me, and I read on. . .

In this book I want to try to do something similar and thus to overcome some of the inaccessibility of science. In so doing I will describe, as a field biologist, the logistics, the peripheral benefits (such as the location itself and the other wildlife), and the thought processes involved in doing science. I will also try to convey something about both the excitement and the frustration of conducting research.

* * *

A close friend once told me, with the slightest hint of envy, that I live a charmed life, and in the sense that my work is also my hobby, this is probably true. I have been fortunate in knowing since I was very small exactly what I wanted to do. For as long as I can remember I have enjoyed watching what birds and other animals do, and many of my colleagues who are professional field biologists were also naturalists when they were young. But what causes naturalists to metamorphose into scientists? Like others interested in natural history, those that are content to remain as birdwatchers are happy simply to record or describe the organisms they see. The scientist, however, wants something more. Through a powerful creative urge they want to generalize, to pull together and synthesize lots of facts and to come up with broad explanations for the patterns they encounter in nature. However, precisely what it is that makes some of us want to 'pull things together', has so far eluded those that study these things. In fact, while there have been attempts to analyse the temperament of scientists who study physics and chemistry (McClelland

1962), I can find no equivalent investigations of biologists. For what it's worth, a study of (male) physicists found that they apparently tend to be loners as children, are 'typically not very interested in girls, marry the first girl they date and thereafter show a rather low level of heterosexual drive'. While it is certainly true that some biologists tend to have solitary childhoods, as for the rest . . . I doubt it.

A multitude of factors determine our future course of development, but some incidents can be particularly potent in shaping our lives. I can vividly recall being taken as part of a family holiday at the age of 11 to Bardsey Island, off the Llyn Peninsula in North Wales. Everything about the place was exquisite: the islandscape, the seabirds, the wheeling flocks of Choughs, and the numerous grey seals. On our walk around the island, we saw someone sitting near the cliff edge with binoculars intently watching some birds and writing down what they saw in a field notebook. As I scrutinized them through my own binoculars, my father suggested to me that I could do something like that in the future, and the seed was sown.

Much later, as a zoology undergraduate my main interests were, not surprisingly, in the behaviour and ecology of animals in the wild. My thoughts became focused during one particular lecture, given by Robin Baker, which described some of Geoff Parker's work on the evolutionary aspects of dungfly courtship and mating behaviour. Parker's study appealed to me because it involved asking very specific questions, couched within a framework of evolutionary ideas. 'Why do male dungflies remain coupled to females after they have finished mating?' I knew from that day that this was the type of biology I wanted to do. Dungflies are not especially charismatic and do not sound very promising as sources of inspiration, but it was not the animals themselves that mattered, but the subject and the way the study was conducted that fired my imagination.

Many of the ideas described in that course of lectures subsequently developed into the discipline now called behavioural ecology. Parker's approach was to test ideas using a common and easily observed organism, which happened to be *Scatophaga stercoraria*, the yellow dungfly. We were privileged as undergraduates to be told about these ideas a year or so before they were published and became sucked into the mainstream of behavioural biology. Behavioural ecology is basically natural history conducted within a robust framework of evolutionary principles (Wilson 1975; Dawkins 1976; Krebs and Davies 1978). Where it fits into the ants' nest of science is shown in Fig. 1. The essence of the behavioural ecology approach is that behaviours, or structures associated with behaviours such as the antlers of deer or the gaudy plumes of birds of paradise, are weighed for their adaptive significance. That is, behavioural ecologists, on seeing an animal with an elaborate structure, or performing a particular behaviour, ask themselves: 'how does this behaviour increase that individual's chances of reproducing or surviving?'.

Much of Parker's research on dungflies centred on their mating behaviour—after all, this is the way that genes are transferred, but I was intrigued to learn that all was not as it appeared. A male dungfly may successfully copulate with a female but still fail to fertilize her eggs. Such reproductive failure can arise

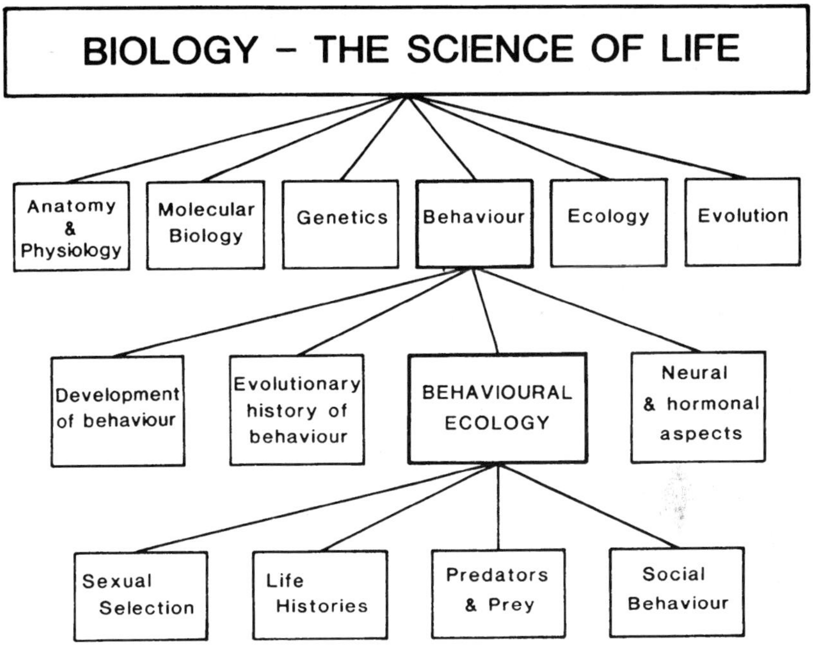

Fig. 1. *Schematic diagram showing the relationship between biology, behaviour, behavioural ecology and some of the different components of the study of behavioural ecology. This figure is not meant to be comprehensive; its aim is simply to illustrate the way ants working on one part of a structure can contribute to the whole.*

from cuckoldry—because another male mates with the same female, displacing the original male's sperm and replacing it with his own! And the reason why males remain attached to females after mating is to deter other males from mating with 'their' female. It is this element of the unexpected that makes behavioural ecology so exciting. Parker referred to the competition between the sperm from different males to fertilize a female's eggs as 'sperm competition' and I wondered if anything similar might occur in birds This question became the major focus of my research over the next 20 years, but as we shall see, by a circuitous, polar route.

In the early 1970s PhD projects for postgraduate students were still very much orientated towards species, rather than being concerned with more general problems, as they are now. My PhD was on Common Guillemots on Skomer Island, off the south Wales coast, and the main reason for conducting this study was that the number of Common Guillemots in Britain had declined dramatically during the previous 40 or 50 years. My remit was to try and find out the cause of the decline. However, there was sufficient scope within the study to try to apply some of Parker's ideas, albeit in a very limited way. The opportunity was limited because the behavioural ecology approach centres on what individuals do and I therefore needed to be able to recognize individual birds, which was not possible, at least not on the scale I required to make sense of their promiscuous mating relationships. My efforts were also limited for another reason. The hardest part of any study comes in deciding exactly which questions one wants to ask, and at that stage I was still very uncertain about just what were the right questions to ask about sperm competition in birds. Once that initial barrier is overcome, however, the rest is relatively easy. Doing science is rather like being dumped on the moors and told to find your way home: difficult, until you are given a compass and a bearing.

* * *

Bardsey was my first island experience: a haven for wildlife, but also the final resting place for large numbers of medieval holy men or 'saints'. Their abundant remains are now occasionally unearthed by the elements and the plough. Like several other islands around the coast of the British Isles, Bardsey held a special place in the hearts and minds of saints, who chose to rest their earthly bodies there. Presumably, by being buried somewhere as beautiful as Bardsey they felt as though they were already on their way to heaven.

For both scientific and other reasons islands have always attracted biologists, such as Charles Darwin, Frank Fraser Darling, Ronald Lockley and others. Scientifically, islands offer the biologist a number of special advantages, one of the most important of which is the finite nature of the animal or plant populations they contain. The island itself sets the limit for the extent of the population; because of the surrounding sea the population cannot extend beyond the island's physical limits. Another feature of islands that attracts biologists is the occurrence, on some islands at least, of unique forms of mainland species. St. Helena in the South Atlantic, for example, boasts the biggest earwig in the world, some 8 cm long! Skomer Island has its own vole, a larger, redder and tamer version of the mainland bank vole, with an unusual tooth (bearing an extra cusp). These special island forms evolve to be different from their mainland relatives for a number of different reasons, but partly because the ecological conditions on islands differ from those elsewhere. The tameness of the Skomer Vole, for example, is at first sight extraordinary: if you catch one and place it in the palm of your hand, it will simply remain there. Try that with a mainland bank vole and you will end up with bloody fingers and the blurred image of a rapidly disappearing small mammal. The fact is, there are no ground-living predators, like red foxes, stoats or weasels, on Skomer so the voles have lost their fear of terrestrial beings, including people, because they pose no threat.

In addition to their scientific advantages, islands also have a special feel to them, and you either like that feeling or you do not. Islands are, by definition, difficult to get to, both for ourselves, but also for many other animals and plants. In a way their very inaccessibility makes them attractive: the unattainable being the pinnacle of desire. In his book *Dream Island*, Ronald Lockley (1930) describes how he lived on Skokholm Island, Skomer's neighbour, in the late 1920s and 1930s. One chapter, entitled 'Possession', describes his acquisition of Skokholm, and this chapter title epitomises one of the sensations that islands engender.

Islands are at once both exclusive and inclusive. The problem of access makes them exclusive in the sense that they exclude others and elicit a sense of elitism because relatively few people have the privilege of living or staying on islands. Waiting to get onto an island because of bad weather, or for whatever reason, produces overwhelming sensations of being excluded. At the same time islands are inclusive because once there you are limited to whatever resources are present on the island: you cannot simply go off to the shops to get something you have forgotten. Most of my feelings towards islands are positive, but I can appreciate those few people that suffer from islandphobia and the dread of not being able to get away. An important aspect of my positive feelings about islands stems from the fact that I, like other biologists, usually go to them with very specific objectives. Without these, islands could readily become rather prison-like. Indeed, for obvious reasons, many islands have been used for precisely this purpose.

Skomer Island off the south coast of Wales. (Photo: N. Alexander).

Of course, islands vary enormously in size and in their distance from the mainland. For biologists to benefit from using them as study areas, islands must not be too large, nor too close to the mainland. Size is probably important for several reasons. First, the larger an island is the greater diversity of habitats it contains, and the more like the mainland it becomes. Second, the larger an island is the less intimately you can get to know it, and the less certain you are about its exclusivity.

There is one further, more sinister reason why islands are important, and why biologists should try to understand them. It is simply this: our entire existence will ultimately depend upon it. As a race we are transforming our world into one huge archipelago of habitat islands, tiny fragments of what once was. Learning to understand the consequences of this process, is just one of the many things we must do to ensure that our own species has any kind of future.

* * *

After completing my PhD I was fortunate enough to get a post as a lecturer in behaviour and ecology at Sheffield University in northern England, where I have remained ever since. Being a lecturer requires one to be a jack of several different trades and a master of as many as possible. In Britain a university lecturer has three main duties: (1) to teach students, both undergraduates and postgraduates, (2) to conduct one's research, and (3) to do some administration. I enjoyed and tried to master the first two, but developed what my colleagues refer to as a cavalier attitude to the third.

The choice of research topic is entirely one's own, and the sense of freedom that this gave me was exhilarating. At first sight it might seem that this liberty provides the perfect opportunity to fritter one's time away on some trivial but pleasurable, research topic. In fact, the choice of research area is crucial since it has to be considered to be worth funding by those bodies that support research, and it must also have some scientific credibility with one's colleagues. A research project whose stated aim was to investigate the link between the wildlife in national parks and the history of art, would certainly be enjoyable, involving as it would visits to the World's major national parks and art galleries, but it would be unlikely to be funded and would leave most of one's colleagues wondering which pigeon hole they should put you in.

My PhD research had prepared me for two possible pigeon holes: either as a seabird biologist, or as a behavioural ecologist. In terms of career trajectories I knew which one I *should* take, but the freedom of choice was too tempting. As the anthropologist Nigel Barley (1988) has so aptly noted, 'The saddest thing about academic research is that when you are young you have plenty of time but no one will give you any money. By the time you have worked your way up the hierarchy, you can normally persuade someone to fund you but you never have enough time to do anything important.' Although I was not really aware of this situation when I started, I can now see all too clearly that it is true, and I therefore have no regrets whatsoever about deciding to take the opportunity that was presented to me: to study marine birds in various parts of the Arctic.

* * *

The Arctic is a great wilderness in the sense that there are few people there, and it is therefore one of the few places left on earth where one can observe wildlife relatively unaffected by human activities. The diversity of Arctic wildlife is limited, but this is more than compensated for by its often incredible abundance, with huge migrating herds of caribou, vast numbers of cetaceans, and gigantic seabird colonies. The Arctic lies at the margin of the universe and we cannot help but be impressed by the ability of animals, plants and native people to thrive under such apparently adverse conditions. The history of the Arctic is one of romance and ego, and from the time of its discovery the harshness of the Arctic has been closely bound up with status-enhancement. Our perception of environmental harshness is one which increases from south to north, and as the title of Nansen's (1897) two volume work, *Farthest North* indicates, the closer one got to the pole, the greater one's eminence. As I was to discover, the same is still true among those who currently work and live in the Arctic. This is not to say that I was personally immune to the glamour of Arctic research, but in this respect I think I was more interested in the way I perceived myself, rather than the way others saw me. Anyway, this was only a small part of my motivation for wanting to work in the Arctic.

* * *

Much of what this book is about concerns birds in the Canadian Arctic. Unfortunately, some of the names that North Americans use for particular seabird species differ from those commonly in use in Britain, despite the fact that most of the species I discuss occur on both sides of the North Atlantic. I had to make a decision about which names I should use, and decided to use the English names. So when I write Common Guillemot, North Americans can

read this as Common Murre, and similarly, Brünnich's Guillemot is the Thick-billed Murre (see Chapter 7). I hope my North American colleagues, and others there who read this, will bear with me. To avoid ambiguity I have used the full names each time; where I refer to 'guillemots' I mean both species. Since there is only one species of puffin, the Atlantic Puffin, breeding in the North Atlantic, for simplicity I refer to this species simply as the Puffin. For other species I give the English name followed by the North American one in parentheses. However, I have followed the Canadian system of presenting all measurements of birds, distances and altitudes in metric units.

Since I have tried to write this for non-scientists I have kept the usual statistics that often appear in books in this series to an absolute minimum. I have tried to make it clear throughout the book when I am discussing something that I, or someone else, has verified scientifically, and I have included references in the standard way so that anyone who wants to can explore these topics in more detail. The text ranges over a variety of subjects, including some non-scientific ones, and I have included references to those too wherever I have thought it useful or necessary, but I should point out that I have not tried to provide a comprehensive coverage of any one topic.

This book is not for my scientific friends. If they read it I hope they might find something of interest in it, but it is not meant for them. Rather, I have written this for those who enjoy natural history and wild places, and who want to know something of what motivates scientists, how they go about their work and what they have discovered. I have tried to write a book that can be read at several different levels. At one, I would like to think it can simply be dipped into, and chapters read more or less in isolation, to get a feel for research and for the Arctic. At a second level, I hope it also provides a readable account of the results of that research. Finally, it provides a personal narrative of seven exciting summers spent in some of the most breathtakingly beautiful areas on earth.

CHAPTER 2

Margins of the universe

After flying for hours over vast areas of ice and bleak terrain the plane started its descent into Resolute. As we approached this Arctic capital from over the sea-ice, I saw the wreckage of several planes, both large and small, strewn either side of the gravel runway. There was a moment of apprehension, but before I had time to think about the implications, we had landed. After standing rather uselessly in the shed they called the airport, uncertain about how I was supposed to manage the next leg of my journey to Pond Inlet, I was identified and taken to PetroCanada's expediter. I was not really sure what an expediter was, but whatever it was, this particular one looked extremely unpromising. Bearing 2 day's stubble on his chin, most of his shirt hanging out, and wearing a pair of worn bedroom slippers, he shuffled out to greet me in an unenthusiastic way. He responded in monosyllables to my questions, and I was momentarily engulfed in gloom: this was not what I was expecting at all. After half an hour or so we left his overheated, undertidied 'unit', and he took me over to the Polar Shelf base for a meal, having said nothing about how I was going to get to Pond Inlet, telling me only 'not to worry'.

I had barely sorted out the eating arrangements and started my meal, when he reappeared and, with a sense of lethargic urgency, told me that there was a flight to Nanisivik in 5 minutes time. We left my uneaten meal and he drove me across the base to where a DC3 Dakota, was standing, with its engines roaring. Not certain where Nanisivik was, I asked him, as I climbed into the aircraft, what I was supposed to do when I got there: he simply said 'Follow the crowd', slammed the door behind me and I was off.

The 'crowd' comprised three other passengers, and unlikely passengers at that. I guessed that they were oil or mineral tycoons since they were dressed in three-piece suits, and one at least was grossly overweight.

The flight was spectacular, and in the clear, low evening light I was able to identify most of the land masses as we flew diagonally across Lancaster Sound. An hour or so later we landed at Nanisivik some 30 km east of Artic Bay on the Brodeur Peninsula. At the time I had little idea of where we were because Nanisivik appears on very few maps. There are good reasons for this. Although I saw no sign of any settlement I later learnt that Nanisivik is a community of 360 non-native people associated with a lead–zinc mine which had been developed for development's sake alone (Brody 1987).

We were on a gently sloping hill and all I could see was the red earth of the runway, highly reminiscent of a building site. Barely before the plane had stopped, the fat tycoon was opening the door. He jumped out onto the 'runway', only to sink up to his calves in red mud. The other two then tried to haul him back into the plane, but it was like trying to haul a bull walrus out of an ice-hole, and their efforts were futile. As they struggled, a truck appeared, reversed up to the plane and we all climbed from the plane onto the back of the truck. I was beginning to wonder where this would take us, when a small helicopter appeared over the nearby horizon and landed within a few metres of the truck. The only occupant leant out of the cockpit and shouted 'Are you Birkhead?', I confirmed that I was, and he told me to 'jump in'. This was sheer magic: out of the mud and into the air and speeding towards Pond Inlet.

Flying relatively slowly at an altitude of just few hundred metres and with all-round vision from the helicopter's bubble the journey was exhilarating: mountains gouged by ice and snow, deep fjords covered with ice and with glimpses of distant birds, mainly Fulmars and skuas, skimming over the patches of open water. At Pond Inlet I was dropped off and met by another PetroCanada expediter, but this one was completely different: human. He took me down to the Pond Inlet hotel and introduced me to the other biologists staying there, including Larry Patterson with whom I would be working at Cape Hay.

* * *

This was my first visit to the Arctic. I was here on an exploratory basis: I wanted to see something of the Arctic and especially to see Brünnich's Guillemots. I had spent the previous five summers studying Common Guillemots on Skomer Island, and was now keen to make some comparisons between them and their northern counterpart. 'Officially' my trip had been arranged through the Canadian Wildlife Service in the hope that I might be able to help Larry with his study and he was here as an employee of LGL, an environmental consultancy. The Brünnich's Guillemot colony at Cape Hay was known to be vast, stretching along some 4 km of 300 m high sheer, limestone cliffs. Even so it had been visited only once before by a biologist, some 30 years previously. The colony occupies a key site at the entrance to Lancaster Sound (Fig. 2), a region renowned for its high concentration of wildlife, and at that stage about

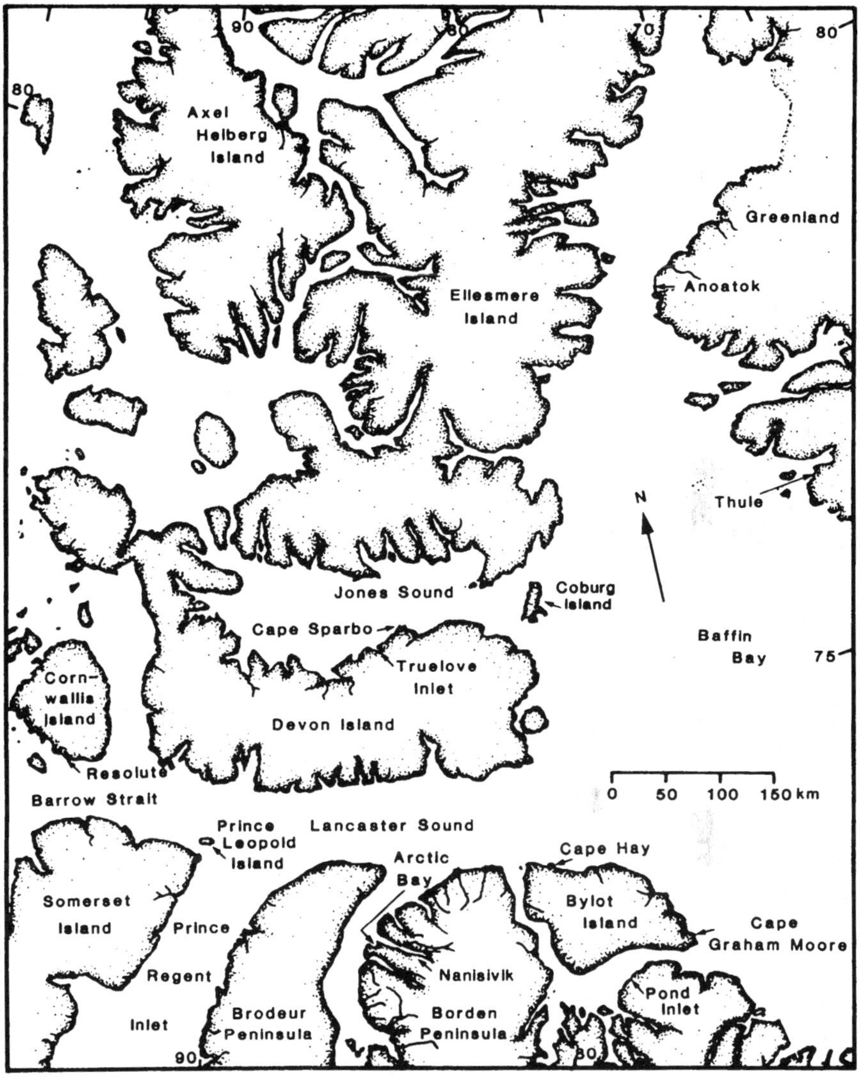

Fig. 2. *Map of the Lancaster and Jones Sounds regions of the northwest Territories in the Canadian Arctic showing locations mentioned in the text.*

to be probed for its oil and gas reserves. By Canadian law even exploratory drilling has to be preceded by wildlife impact assessments, and to that end LGL were about to start monitoring work at Cape Hay and elsewhere.

* * *

The dining room of the Pond Inlet Hotel was full of wildlife biologists of one sort or another. Most of them were young, but a few had nevertheless spent several summers at high latitudes. As I sat down next to Larry I found myself listening to a conversation about narwhals. A group of biologists interested in the diet of these little-known cetaceans were taking advantage of the opportunity provided by the local Inuit, who were engaged in their traditional hunt at the ice edge. Their study involved camping on the sea-ice, next to the Inuit and close to where the ice meets the sea. The narwhals migrating into Lancaster Sound from the south, followed the ice-edge, and were thus easy targets. In times past they would have been hunted using a harpoon and the battle between man and whale would have been fairly even. But now, as the biologists revealed, it was a different story. Armed with .303 rifles the Inuit turned the hunt into a horror story. As the whales surfaced for air they were shot at almost point blank range. Many were hit, but in sharp contrast to the days of the traditional harpoon, many now sank without trace. Of course, there are strict quotas on how many narwhals the Inuit can kill. But those whose tusk was too short or that died on the sea bed or beneath the ice were not counted, and the biologists reckoned that only 1 in 4 was recovered. Once a whale was killed and on the ice, the biologists were able to work and examine the whale's last supper.

Only the male narwhal possesses the long spiralled tusk, which is a single elongated left incisor tooth, and they use it in their competition for females. They have been seen jousting with their tusks projecting above the water, and only males bear the scars from these tournaments. In the middle ages the narwhal's tusk, which may be over 2 meters long, was passed off as the horn of the unicorn, imbued with magical powers. Even today narwhal tusks still have a deadly fascination and most tusks are sold to wealthy businessmen with flagging egos. I had seen tusks for sale in Frobisher Bay on my way north, and at that time they were selling at about $50 per lb and $750 for a 2 m long tusk. The lucky Inuit who killed a rare 'double tusker', which make up about 2% of the population, could get more from that one skull than he would normally do from a full season of hunting. The final irony was that little or none of the slaughtered narwhal's carcass was eaten: some of the skin, referred to as mukluk, was taken, but after the head was removed the rest of the carcass was abandoned.

Larry told me about his efforts to get to Cape Hay to start the Brünnich's Guillemot study. Two weeks previously a helicopter had got them and their equipment out to the colony, but they had been forced to return to Pond Inlet because they had not yet received official permission to be there. The plan was that they would commute on a daily basis until permission was granted. I mentally rolled my eyes. However, Larry told me, it had not mattered, because the fog was so dense at Cape Hay that despite several attempts it had not been possible for a helicopter to get in anyway.

* * *

I prepared myself for a long stay in the Pond Inlet Hotel, and decided to

A male narwhal killed by Inuit hunters. (Photo: F. Breummer).

explore the surrounding tundra. From the air, Pond, as it was referred to, had looked beautiful, but from the ground the wrinkles and blemishes were rather more obvious. The entire shoreline was strewn with Inuit garbage, both biological and abiological. The latter comprised mainly plastic refuse and cans, but the biological material was more interesting and included the washed-up carcasses of narwhal and belugas, shot birds and the huge vertebrae of long-dead whales. The vertebrae of whales seem to be unique in having plates of bone on their articulating surfaces, which become detached in long-dead animals, forming circular, biscuit-shaped discs.

There has been an Inuit settlement at Pond Inlet for centuries, but it had existed in its present form of a hundred or so prefabricated, southern-style houses, only since the late 1950s when the then Canadian government decided it would be a good thing (for whom?) if the Inuit ceased to be nomadic but were settled more permanently (Brody 1975). One of the negative consequences of a permanent settlement is the accumulation of refuse. Combined with the unthinking imposition of western consumerism, the result was a society in danger of being suffocated by its own dirt. Everywhere in Pond, in the hotel and the muddy streets, the air had a heavy, distinctive fetor. The stench was that of the ubiquitous 'honey bag'. The permafrost prevents any kind of sewage system, and so all toilets simply collect faeces into green polythene bags, which when full, or preferably when only part full, are dumped outside, adjacent to the road, where they are duly collected by the honey bag truck. The truck then carries them all to the dump, about 2 km from the town where they are abandoned to the ravages of weather and dogs. The year-round low temperatures mean that there is little bacterial decomposition, so the contributions of some 700 people mean that the dump gets bigger and bigger, and the smell is simply accepted as part of the town environment. I visited the dump to witness this sorry spectacle for myself: a mixture of honey bags, rotting food, burning furniture and the worn-out corpses of skidoos and huskies. I looked with disbelief, and wondered whether this enormous, green amoeba-like organism would ever pick up enough momentum to carry itself down the hill into the sea. Less concerned than I was, a few Snow Buntings and Lapland Buntings (longspurs) foraged for flies around the edge of the dump.

Away from the settlement the air was fresh and unbelievably clear. Despite the warnings I had heard about the rigours of the Arctic, the temperatures were no different from those during a normal British winter, and certainly considerably higher than those of a winter further south in Canada. As we waited for the fog to clear at Cape Hay, Larry and I used the time to explore. We walked along the shore west of Pond, to visit the site of an ancient Thule eskimos settlement, still in a remarkable state of preservation after several hundred years. This must have been a perfect location for a camp. It was situated on a spit of land above the shore beside a small river, with magnificent views of the sound across to Bylot Island, to spot the bowhead whales migrating into Lancaster Sound for the summer. Their houses were made largely from bowhead bones: the huge skulls and vertebrae, carefully covered with earth sods forming the circular base, and the ribs used to form a roof that would have been covered with skins. Over the centuries the bones had fertilized the ground and in stark contrast to elsewhere, the vegetation around the Thule houses was green and lush. The difference from the other settlement, just a mile or so away, was striking. Inside one of the houses I found the skull of a husky with grass growing up through it, which, paradoxically, gave the place an even greater sense of life.

In fact the contrast between life and death was everywhere here. Walking back towards Pond, I was shown the graves of two early pioneers. The graves were shallow and covered only by a sheet of plywood. On looking inside I was amazed to see both skeletons still wearing most of their clothes! Larry repeated the (apocryphal?) story he had been told: that they had been found frozen to death in their cabin. My train of thought was broken by the sound, and then the sight of a helicopter heading away from Pond towards Cape Hay. We rushed back to the hotel where we learnt that the mammal team had indeed set off, but that the helicopter would make a second trip for us. But even as we were talking, we received the radio message that Cape Hay was 'socked in' (with thick fog) and the helicopter, unable to land, was returning.

The Arctic is a vast wasteland, studded with occasional oases. On my flight from Resolute to Pond Inlet I couldn't help thinking that most of what I could see was rather desolate: plenty of mud and gravel, but little sign of life. Once one reached an oasis, a location where conditions conspired to make life not only possible, but abundant, the contrast was remarkable. The area around Pond Inlet was like that. If I ignored the exceptional clarity of the air, the sea-ice stretching across to Bylot Island and the views of distant mountains rising sharply from the sea, I could have almost been on the moors near my

Fast ice between Pond Inlet and Bylot Island in the distance.

Sheffield home. This was comparatively rich tundra with extensive plant cover consisting of short scrubby, heath-like vegetation on rolling hills. There were Baird's Sandpipers and Golden Plovers everywhere, and on almost every pool a pair of Grey Phalaropes (Red Phalaropes) pirouetted. Overhead pairs of Ravens cronked deeply to each other.

Despite the novelty of my surroundings I was starting to feel frustrated at being here and not at Cape Hay. After 3 days of waiting I came across what I considered to be a good omen. I had been walking several kilometres from the settlement along the shore, and as I returned at about 23.00 h, in the uncanny brightness of full June daylight, I found, lying in the heath, the skull of an enormous Raven. Animal skulls, and those of birds in particular, are so beautifully sculpted I have always collected them, under the guise of science. The truth is, though, that I just like them for what they are. I had even fixed it so that my undergraduate project had been on skulls: those of the crow family. I had collected hundreds, mainly birds shot by farmers, but since Ravens cannot legally be shot in Britain I had acquired only two of my own and in order to complete my study, I had had to borrow the rest from museums. Ravens have always been regarded as special birds (Heinrich 1990) because of their impressive size, life-long pair-bonds and extraordinary intelligence. They feature regularly in both Inuit and Red Indian folklore and the Vikings apparently used them on their voyages of discovery. After sailing for several days with no sight of land a Raven would be released—if the bird returned to the ship they were forced to continue, but if the Raven flew off in a purposeful manner then the Vikings knew that land lay ahead. The Raven is also one of the few birds that remains in the Arctic throughout the year, apparently unaffected by the perpetual winter darkness and sub-subzero temperatures. I picked up the skull, and after clearing some off some old feathers and remnants of dried skin, I turned it over in my hand, admiring its flawless, potent form. Despite lying there on its own I was surprised and pleased to see that the tiny pterygoid bones, the ones which control the movement of the upper mandible, were still in place. These loosely attached bones are usually the first to be lost from birds' skulls.

* * *

As our helicopter rose over the final ridge of Bylot's mountains early the next morning I could see before us a wide, flat plateau. There were patches of snow and muddy grey ground, but no sign of vegetation: this was a bleak place. I became aware of three tiny shapes disrupting the distant flat horizon, and gradually realised that this was 'camp'. As the chopper came in to land I could see that one of the three tents had collapsed under the snow, but the other two looked okay. At that point the sun was shining and looking back towards the way we had just come the snow-covered peaks provided a magnificent back-drop to an otherwise featureless landscape. There were four of us, Larry and myself studying birds, and Paul and Markusi, the mammal team. Markusi was a Pond Inlet Inuit, a token gesture of local employment. He spoke nothing but

The camp near Cape Hay: three flimsy tents provided little protection from the elements or polar bears.

Inuktituk, but proved to be resourceful and immediately effected a sophisticated repair to the tent using very little but his ingenuity.

Each team had a sleeping tent, and we shared the central tent for cooking and eating. The tents were tall enough to stand up in, and about 3 m long, but the facilities were disappointingly primitive: a single burner stove, no table, no chairs and no ground sheets. I wondered whether I was soft expecting such luxuries, and concluded that real Arctic biologists probably did not worry about such things. As we unpacked I discovered to my relief that there were camp beds, or cots as the others called them, which raised us above the primeval, muddy gravel, at least while we were sleeping.

Larry pointed the way to cliffs and I went off to explore. Snow Buntings hugged the ground in ones and twos, and a pair of Ravens flew past. Although the sun was bright, a dense, surreal mist rose from the ground and in front of me the Brocken Spectre of my image was surrounded by colourful a halo. As I squelched over the uneven terrain, I suddenly became aware that I was almost up to the cliff edge, and rather than coming from the ground the mist was coming from the sea. The effect was utterly bizarre: the mist rose in a solid wall up from the sea several hundred metres below to the level of the top of the cliffs, and then spilled gently over onto the land. By walking right to the sheer edge of the cliffs I could look down the narrow gap between them and the mist onto an icy shore-line. A blast of cool air hit me, together with the distant roar of vast numbers of Brünnich's Guillemots that covered the cliffs. They sounded like harsh, staccato dogs barking at each other. On the snow-covered

Brünnich's Guillemots at Cape Hay.

shelf of ice jutting out from the foot of the cliffs I could make out the footprints of polar bears through my binoculars. I was thankful that they were down there and that we were at the top of the cliffs. As I sat and watched the Brünnich's Guillemots my excitement started to mingle with concern as I wondered how on earth we were going to study birds that were so far away?

I decided to head back to the tents, but on looking up I discovered that the mist had now surmounted the cliffs completely and was rapidly obscuring everything in sight, including the camp. The sun disappeared and the atmosphere changed from one of bright, warm optimism to cool, damp concern. This was Cape Hay. We saw the sun on only one other day during that trip, and the team that came back the following year (Chapter 3) had the same murky experience. Listening to our radio as the other camps gave their weather reports to Polar Continental Shelf at Resolute, it was clear that meteorological conditions at Cape Hay contrived to keep the area of the seabird colony under one large, perpetual lump of fog. Even just 8 kilometres away it was usually clear and sunny.

The two mammal people were at Cape Hay to count the whales and seals migrating into Lancaster Sound for the summer. Under clear conditions the cliffs at the Brünnich's Guillemot colony would provide a wonderful vantage point for counting marine mammals, but as it was Paul and Markusi struggled to see the sea, let alone any marine mammals. Markusi was content however, since he was being paid a hefty salary regardless of whether he saw any whales or not, and he had plenty to occupy his time. Back in Pond he earned his living as a carver, selling his work to the Inuit co-operative who then distributed his carvings to souvenir shops in southern Canada and elsewhere. Indeed, most of

Markusi's luggage comprised lumps of soapstone and the various files and hacksaws that would transform those lumps of cold green stone into the smooth, chunky carvings we now associate with the Inuit.

Markusi did not speak English but he was able to understand most of what we said. Nonetheless, our 'conversations' ended up pretty much like a game of charades. We discovered that his name was the Inuit version of 'Mark', which had been allotted to him by a missionary some 30 or 40 years earlier. His name, with its biblical connotations, seemed to me to embodiment of the 'moral serfdom' imposed on the Inuit by those who should have known better (Brody 1975). On the other hand, the addition of the Inuit 'usi', which effectively destroyed any religious effect the original name might have had, held out a ray of hope.

Our first meal time had all of us feeling slightly ill at ease: Markusi obviously never used a knife and fork, but presumably because I had handed them out, he tried to make a go of it. He was embarrassed at his inability and we were embarrassed at having forced him into this situation. Fortunately the tension snapped when, with a grand gesture, he abandoned the cutlery and resorted to fingers.

We also differed in our approach to cooking. The National Museum of Canada had asked us to obtain some Brünnich's Guillemot study skins for them. Since the museum required only the birds' skin and because I knew that in Newfoundland guillemots were a favourite dish (see Chapter 4), I was interested to discover what they tasted like. After the birds had been skinned I fried the breast meat in garlic and it was delicious, tasting like a mixture of prime steak and liver. But on looking across to Markusi I could see he was not impressed. He indicated that the next day he would show me how to cook properly. I waited with mixed anticipation. The next evening at supper time he went to the entrance of the tent he shared with Paul, and opened a large black garbage bag which was lying outside. Choosing between whale and caribou meat, he pulled out a large lump of the latter and simply set it in a big pan of water to boil. Even by my simple culinary standards this seemed unpromising, yet the end result was excellent: succulent and tasty. Markusi then offered us some of the liquid remaining in the billy can. Prudishly we refused: he shrugged in mild disbelief and drank the litre or so of brown soup with obvious, noisy pleasure.

* * *

The summer of 1978 was a harsh one in the Canadian Arctic. Instead of breaking up and dispersing in May, the ice in Lancaster Sound barely shifted all summer, with the result that conditions for seabirds and other marine animals were much more difficult than in other years. Nowhere was the effect of these severe ice conditions felt so intensely as on Prince Leopold Island, lying at the western end of Lancaster Sound at the junction of Prince Regent Inlet and Barrow Strait (Fig. 2). In the previous 3 years Tony Gaston and David Nettleship of the Canadian Wildlife Service had conducted a detailed study of the breeding biology of Brünnich's Guillemot and other seabirds at

this colony, but they had never recorded ice conditions like those of 1978. Prince Leopold Island was first recorded in August 1819 by William Parry, during his exploration of Lancaster Sound, and he named it in honour of His Royal Highness Prince Leopold Saxe-Coburg. It was not until 139 years later, in August 1958, that a biologist first visited Prince Leopold Island and discovered its ornithological riches. Most seabird colonies in the high Arctic typically contain just one or at most two species, but at Prince Leopold Island there was a particularly rich community. In the late 1970s it comprised Brünnich's Guillemots (86,000 pairs), Fulmars (62,000 pairs) Kittiwakes (29,000 pairs), 4000 pairs of Black Guillemots and about 200 pairs of Glaucous Gulls.

On my way to Pond Inlet I had flown past Prince Leopold Island. Measuring 11 × 8 km, almost oval in outline and bounded by 300 m high cliffs, it looked like a gigantic Christmas cake lying on a tray of icing sugar. Although I could not see them, a group of biologists whom I knew had their camp at the southwestern corner of the island. The main study of Prince Leopold's marine birds had finished at the end of the previous breeding season, but the opportunity to discover what effect these unusual ice conditions would have on the birds was too good to miss, and David Nettleship had managed to get a team of biologists together at short notice. One of the objectives of the original 3-year-long study of seabirds at Prince Leopold Island was to provide information on when the birds bred and how successful they were at rearing chicks.

However, ice and weather conditions had been fairly uniform during those 3 years (Gaston and Nettleship 1981) and it was important to make the most of this opportunity to see how the birds coped under more difficult circumstances. Obviously, the failure of the ice to move out of Lancaster Sound meant that the birds would have to travel much further than usual to find open water in which to feed. Moreover, since feeding areas would be few and far between once the birds got there they would probably face stiff competition from others. Reduced access to food could severely disrupt the birds' breeding cycle and cut down their breeding success. Cape Hay was adjacent to open water where seabirds could feed, and we attempted to collect similar information to that being obtained at Prince Leopold to see if our birds were less severely affected.

In contrast to the team at Prince Leopold Island who had a well appointed camp with well established study areas and routines, I felt that our ability to collect good quality information was seriously constrained. Anyone can be uncomfortable in the field, and being uncomfortable is incompatible with doing good field work. Our camp was extremely basic, and the tent which Larry referred to as the hide, not exactly conducive to the long periods of exacting observation which are necessary to measure breeding success. The

The seabird cliffs near Cape Hay showing the low promontory from where I watched the snow and ice-fall that killed large numbers of Brünnich's Guillemot. The near-vertical snow fields are clearly visible above the cliffs on the right hand side. For some indication of scale the cliffs on the left hand side of the picture are just over 330 m high. (Photo: D. N. Nettleship).

traditional seabird biologist's hide is a stable, wooden structure just large enough to fit snugly into, with a smallish window to provide a clear view of the birds but without letting in too much weather. The Cape Hay hide was a tent large enough to hold a wedding reception. It was perched at the very lip of a sheer cliff, so that should one overbalance the only safety net from infinity was a thin layer of canvas. The 'window' was actually an entire side of the tent, and the whole structure flapped and wobbled in the wind like a stranded fish gasping its last breath. To add insult to injury I was concerned that the group of Brünnich's Guillemots we could watch from the tent was probably completely atypical. It was the only one right at the very top of the cliff, and unlike those further down it was particularly vulnerable to Arctic Foxes.

The effects of a severe food shortage for seabirds could manifest itself in several different ways. One thing I expected to happen was that the extensive ice cover would result in the Brünnich's Guillemots laying relatively small eggs, particularly at Prince Leopold Island. We therefore needed to find some part of the colony where I could measure a sample of eggs and test this idea. It would also mean that I did not have to sit in that awful tent-hide quite as much. However, for most of the length of the colony the cliffs were quite perpendicular, we had no climbing equipment and the birds were situated in horribly unsafe places, so measuring eggs was not going to be easy. Larry and I walked along the cliff-tops in the fog trying to find a suitable place to reach some eggs. The terrain was variable and extreme: in some areas the ground consisted entirely of football-sized boulders and covered in orange lichen (*Caloplaca*). The fog made the lichen incredibly slippery and, as I discovered, our feet are definitely designed for walking on flat, level surfaces rather than such irregular ones. There were the occasional signs of an Inuit presence: we came across several small, crude stone-built fox traps barely distinguishable among the boulders. The traps were covered in *Caloplaca* too, indicating that it was many years since the Inuit had been there. In other areas the frost had heaved and churned the mud into irregular polygons, along the edge of which a scant line of vegetation grew. The plants were mostly purple saxifrage, and there were also a few mosses and lichens, but it was not what one would call a luxurious growth.

At one point towards the western end of the colony, a large snow field sloped down to the cliffs and fed a small stream which pitched itself over the cliff edge only to be destroyed by the vigorous cool updraft. The sheets of spray froze onto the vertical rocks in ever increasing grotesque shapes, hanging ominously over the Brünnich's Guillemots stolidly incubating their egg. The birds seemed to be unperturbed by the ice and the constant splash from the nearby waterfall. However, when a Peregrine Falcon suddenly swept along the cliff-tops their reaction was immediate: they hurled themselves off their breeding ledges, dropping vertically like so many rocks, before opening their wings and sweeping away from the cliffs only a few metres above the sea. I suppose if you are a guillemot this is probably the best way to avoid being taken by a raptor. Interestingly, in other colonies where Peregrines or Gyrfalcons breed almost amongst them, Brünnich's Guillemots rapidly get used to the raptors' presence (Gaston *et al.* 1985).

Eventually we found a place where measuring a sample of eggs was just possible. At the far western end of the colony the snow field descended several hundred metres in a steep slope to a small promontory about only 70 m above the sea. Beside it was a deep bay, whose ledges were packed with thousands of Brünnich's Guillemots and Kittiwakes. At last we were reasonably close and at the same level as the birds. As I scanned the ledges with my binoculars, searching for potential study areas my eyes settled on a curiously pale bird amongst the Brünnich's Guillemots and I realized it was an albino—the first one I had seen. But my attention was snatched away by an bewildering, unearthly roar coming from above the cliffs. I could not understand where the noise was coming from, but the sound sent adrenaline surging through my bloodstream. A huge area of frozen snow and ice on the slopes above the birds had detached itself and was slowly sliding towards the cliff edge. I watched with disbelief as the snow and ice knocked dozens of birds from their ledges. Some managed to take flight amidst the avalanche, while others simply disappeared under the tonnes of falling snow and ice. The birds that remained on their eggs were completely buried and afterwards those that were still alive struggled to extricate themselves. I crept forward to the cliff edge to see what had become of those birds knocked onto the ice below. Many were buried under the pile of snow but others were crippled and helplessly flailing around on the landfast ice. Within seconds the ever resourceful Glaucous Gulls were there, tearing into the Brünnich's Guillemots, both dead and alive. I recovered

my breath, and realized just how different this environment was from Skomer Island, both for seabirds and biologists.

Notwithstanding, this part of the colony proved to be both productive and a relatively pleasant place to work. Often we were actually below the fog, and there were plenty of birds breeding in very accessible sites. On one occasion as we made our way down to the peninsula we could see a group of 16 Narwhals, males and females with calves, lying together in bodily contact in the water just under the cliffs. The world population of Narwhal is small, somewhere around 20,000 individuals, and 85% of these spend their summer in Lancaster Sound. Watching these few individuals through binoculars we admired their beautiful grey-flecked skin and the extraordinary tusks of the males. The peninsula had also been a place that the Inuit had favoured, perhaps for hunting Brünnich's Guillemots, since we found a crude stone shelter, constructed in such a way that using a long handled net they could have caught birds flying overhead. From our point of view too the peninsula was perfect: it allowed us to take Brünnich's Guillemots off their breeding ledges in order to weigh them and to measure and weigh their eggs. Anxious to start collecting some useful information I edged forward onto the cliff ledge. The first bird I encountered simply allowed me to remove its egg from beneath it and after I had done what was necessary, allowed me to return it. Other Brünnich's Guillemots were less obliging and left the ledges as I approached, but after a few minutes they returned, and stood close beside me as I worked.

* * *

The fog persisted day after depressing day. Often it was so bad that we would lose our way just going out to get drinking water from one of the nearby melt pools. I once spent an hour wandering round looking for the tents after collecting some water. After that I was much more careful. We built a series of inukshuks ('stone men') about 20 m apart between the camp and the cliff-top to help us find our way, but sometimes even these were not good enough, so we topped each one with red fluorescent tape. Hardly aesthetic, but effective.

Then one morning we awoke to find the tent full of an unfamiliar bright light. Outside we could hardly believe our eyes: perfectly clear blue skies! It is difficult to describe the effect of this. After living in a dim, murky fog since we arrived, and then suddenly having clear weather was as though we had been transported overnight to a new location. For the first time we could see along the spectacular length of the cliffs, ultimately to Cape Hay itself, 8 km to the east. To the north looking 100 km across the full extent of the sea-ice choking Lancaster Sound were the icy peaks of Devon Island.

I decided to use the day to get a good feel for the colony, and set off eastwards with my lunch, binoculars and note book. One hundred years earlier, in 1858, Leopold M'Clintock, who was searching for the lost Franklin expedition, had been told of a shipwreck near Cape Hay and came to investigate. He recorded the presence of the Brünnich's Guillemot colony in his journal, but noted that the extensive sea-ice prevented him from 'levying a tax' upon the colony. The first non-Inuits to visit Cape Hay were Les Tuck, the

The cliffs near Cape Hay on a clear day.

pioneer of guillemot biology, and RCMP corporal Ray Johnson; they were accompanied by Mucktar, an Inuit from Pond Inlet. They had come here in 1957, travelling by dog sled over the sea-ice from Pond Inlet. Johnson returned just before the sea ice started to break up, while Tuck and Mucktar spent the summer counting the huge number of birds breeding in this colony. Unlike us they had worked mainly from ice- and sea level.

Tuck estimated that there was a total of 400,000 pairs of Brünnich's Guillemots breeding along the 4 or 5 km of vertical cliffs. My experience of counting Common Guillemots had been limited to small colonies, and as I made my way along the colony on that bright sunny morning, I was over-

The Brünnich's Guillemot cliffs near Cape Hay. The person standing on the cliff top (upper left) gives some idea of the scale.

whelmed by the number of birds, the scenery and Tuck's achievement. The problem of counting the birds, which was such a crucial part of their conservation, required an entirely different scale of thought from that which I had been used to. In fact, it was not part of our objective to try to count this colony, but thinking about it was a salutary exercise.

I walked to the eastern end of the colony, seeing as I did so numerous Snow Buntings and a small party of brilliant orange Knot. I slithered down a long scree slope, past a rocket-shaped tower covered in breeding Brünnich's Guillemots, onto a large open area of flat ground. When I was several kilometres from camp, I looked down and saw in the mud a single large footprint of a polar bear. My first thought was as to how recent this was: if it was extremely recent I had probably had it, since I had not brought a gun and there was absolutely nowhere to run to. I had a good look round, and extremely vigilant I carried on. Close to where I had seen the footprint I found the skull of a harp seal—presumably killed by a polar bear judging from the way it was damaged.

Also, it was difficult to imagine what else would have carried a seal so far from the sea. After a while I stopped thinking about the bear, and in fact, until that first close encounter with polar bear (or any large predator), one assumes (foolishly) that one is immortal.

Coming over a small rise I was surprised to hear the high pitched barking of an arctic fox. Looking through the binoculars I could see two well-grown fox cubs standing beside a large boulder. As I approached they both disappeared into a cavity beneath the rock. The area around the den was littered with the remains of a large number of Brünnich's Guillemots, presumably the scavenged bodies of birds killed or injured by ice and rock falls. The remains were all identical: a pair of wings held together in the middle by the pectoral girdle and a piece of sternum. I retreated a few metres and made a kissing sound: the foxes immediately reappeared from the den, their heads cocked to one side. Standing just a metre away they stared at me and tried to work out who was making this unusual noise. This was a small litter by arctic fox standards, since as many 10 cubs are occasionally born. The number of young produced depends upon food supplies, and I imagined that the adult foxes had probably had to work quite hard to collect the bird corpses. Arctic foxes usually breed as monogamous pairs, but they sometimes have a more complex breeding system with an extra female (who is usually one of the pair's offspring from a previous year) acting as a helper (Hersteinsson 1984). In this particular case I saw only single adults at the den at any one time.

Leaving the foxes I wandered further along the coast towards Cape Hay itself until I came across two tent rings: circles of stones used to hold down the sides of tents. At the time I assumed that this was the site of an Inuit camp, and probably a very old one since there were no modern artefacts such as empty cans, anywhere. Subsequently I wondered whether this was where Les Tuck and Mucktar had camped 20 or so years previously.

I made my way back to camp to find that the mammal watchers had had their first successful day, with large numbers of narwhal and harp seals swimming west into Lancaster Sound. We ate supper outside in the still brilliant sunshine, and afterwards Markusi brought out a small carving that he was working on. I was pleasantly surprised to see that it was a stylized but attractive Brünnich's Guillemot. I failed to put two and two together and Larry helpfully suggested that Markusi had probably made it for me. Using our sign language, I asked him if he wanted to sell it. He indicated that he still had to finish it off, but once he had completed it $10.00 would be fine. I agreed and we were both delighted with the arrangement. Although the large Inuit soapstone carvings of things like grappling polar bears or copulating walruses that one sees on sale in Canadian souvenir shops are impressive, they are atypical. Traditionally the Inuit carved animals from walrus ivory or from soapstone for their children. These toys were always tiny because, prior to their being tied to settlements, the Inuit's nomadic lifestyle meant that everything had to be portable.

It was midnight and the sun continued to burn brilliantly; I was tired, but reluctant to go to bed while the weather was so good. I walked back to the cliffs to the end of the colony to look down upon the area of the fox den. As I reached

Markusi, an Inuit field assistant, putting the finishing touches to a soapstone carving of a Brünnich's Guillemot.

the top of the scree slope I could see an adult fox and several well-grown cubs far below, racing backwards and forwards in what seemed like a game of tag. Through the binoculars I could see that something rather more serious was going on. One of the adult foxes had caught a white arctic hare and was trying to cache it, using its nose to bury the hare under mud and rubble. As soon as the hare was buried one of the pups would race back to unearth it and run off with it. The adult fox then chased the offending pup, snapped at it and regained the now muddy hare corpse, and started again. I watched the foxes for an hour or so during which the performance was repeated several times. I began to feel a dampness in the air, and turning round I saw to my dismay that the fog was returning, rolling up the vertical cliffs from the sea, obscuring the sun, and floating like a gigantic medusa towards our camp.

* * *

Fig. 3 *Map showing the location of major Common and Brünnich's Guillemot colonies in the eastern Canadian Arctic. 1. Coburg Island, 2. Prince Leopold Island, 3. Cape Hay, 4. Cape Graham Moore, 5. Reid Bay, 6. Hantzsch Island, 7. Akpatok Island, 8. Digges Island, 9. Coats Island, 10. Gannet Islands, 11. Funk Island, 12. Witless Bay (Great, Green and Gull Islands), 13. Cape St. Mary's. Solid symbols indicate Brünnich's Guillemot colonies, split symbols indicate mixed Common and Brünnich's Guillemot colonies. There are no major colonies of Common Guillemots without some Brünnich's Guillemots in this region. From Nettleship (in press).*

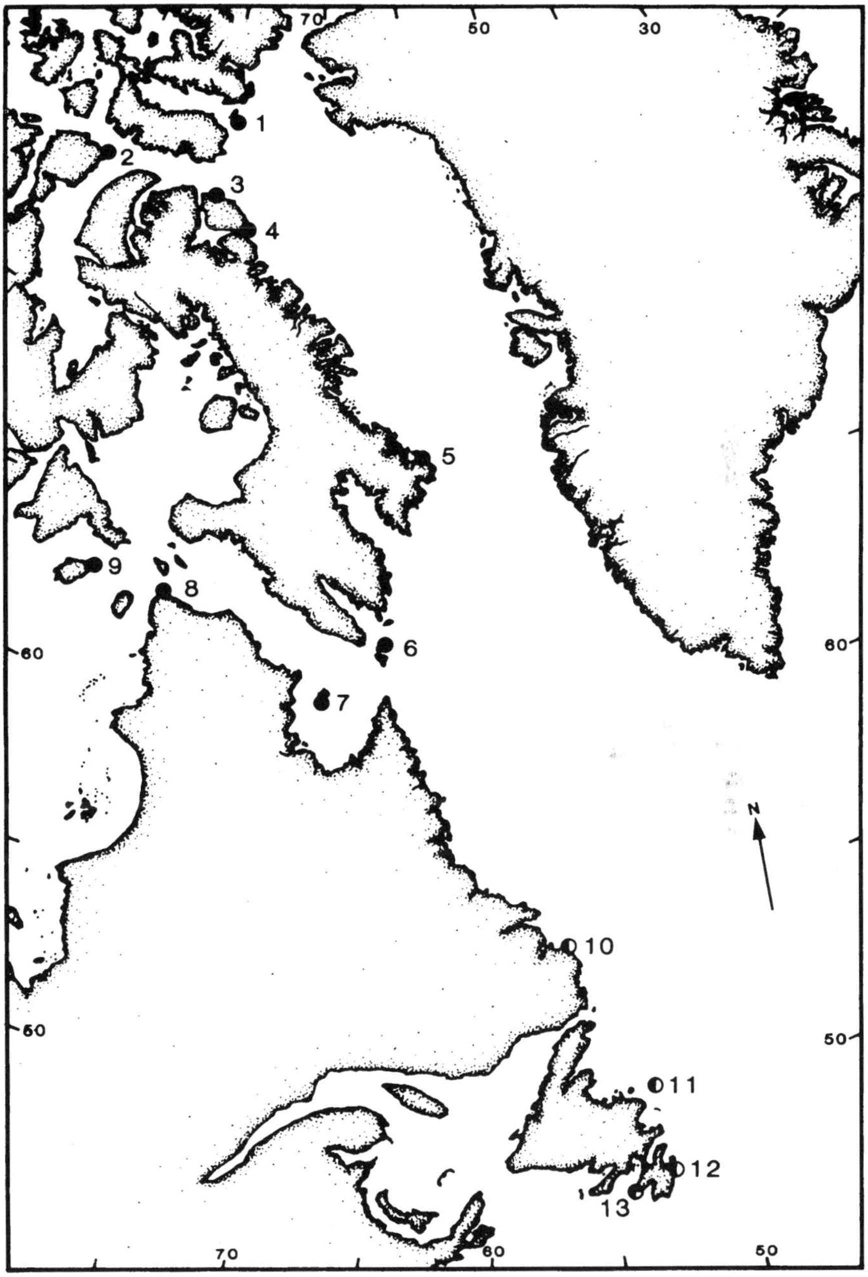
70
50
30
1
2
3
4
5
6
7
8
9
10
11
12
13
60
60
50
50
N
70
60
50

Lancaster Sound remained choked with ice throughout the summer, and as we had expected, the seabirds at Prince Leopold Island suffered as a consequence. The difficulty the birds had in finding sufficient food expressed itself in different ways, according to each species' feeding ecology. Many of the Kittiwakes and Fulmars did not even bother to attempt to breed, but most of the Brünnich's Guillemots did, albeit considerably later than in the previous years. Because Brünnich's Guillemots feed below the surface, often diving to considerable depths (see Chapter 6), they appeared to be less affected by the ice cover, having a much greater volume of water in which to search for food than the surface-feeding Fulmars and Kittiwakes. However, the eggs the Brunnich's Guillemots laid that year were still about 10% smaller on average than those in the previous 3 years.

In this particular year the ice reduced the potential feeding area for the surface feeders so much that it simply was not worth the effort of laying an egg, presumably because there would be insufficent food to rear any chick that hatched. For the Brünnich's Guillemots, on the other hand, it was worth risking a breeding attempt. However, none of us, and certainly no seabird, can see into the future and these decisions are based on what is most likely to occur. As it turned out, the Brünnich's Guillemots' breeding season ended in disaster.

Brünnich's Guillemot with a sculpin. (Photo: E. Greene).

Most Brünnich's Guillemots laid an egg and managed to incubate it successfully. Their chicks were slightly undersized at hatching, and because the parents fed them so infrequently, they did not grow quite as rapidly as they would have done when food been more abundant. A slow growth rate in itself was not critical because the chicks of Brünnich's Guillemots seem to be remarkably flexible in this respect. The crunch came when the young birds were due to leave the colony. Typically, young guillemots (and Razorbills), 'fledge' when only one-quarter adult size (see Chapter 6). They are still flightless at this stage, but through a mixture of parachuting and gliding they get down from their lofty natal ledge onto the sea. Once there, their father greets them, and the two swim out to sea. The birds from Prince Leopold Island would normally use the currents to carry them east into Baffin Bay, then on down to west Greenland and eventually south into Newfoundland waters (Gaston 1980). However, in 1978 there was virtually no open water for the young Prince Leopold Guillemots to alight on as they plunged from the vertical cliffs. Instead, most of them landed on the ice, and remarkably, they survived what must have been an unpleasant landing. The male parents also landed on the ice alongside them, and the pairs of fathers and offspring then started the long, long walk towards open water over 60 km away, near Cape Hay.

The biologists I spoke to at Prince Leopold Island told me of this heartrending spectacle. They watched helplessly from the cliff tops as 50,000 or so chicks and their male parents formed a doomed, black cortege, making its way like a crippled snake across the sea-ice into the distance. The Glaucous Gulls and arctic foxes had a feast, taking both the exhausted chicks and the adults who, having committed themselves to walking across the ice, were unable to take flight and were completely helpless when confronted by these predators. As my colleagues described these events to me their normal scientific objectivity gave way to emotion, and tears welled up in their eyes.

It seems likely that none of the Brünnich's Guillemot chicks from Prince Leopold Island survived that year. After its chick was dead the male parent that had successfully run the gauntlet would have had a chance to fly south once it found one of the open leads between the ice pans. Although this incident was depressing, it was the result of natural variation in climatic conditions, and in the long-run, probably had no serious effect on the total breeding population of Brünnich's Guillemots of Prince Leopold Island. How could 50,000 chicks die, with no effect? Most seabird species experience catastrophic breeding seasons, and as many as 1 in 10 seasons probably result in breeding failure (Wooler *et al.* 1992). However, the longevity of the adults means that the impact of such seasons is spread over many years, and is thus diluted for the population as a whole. In addition, the death of all the chicks in 1978 may have resulted in there being more food and better survival for the adult Brünnich's Guillemots that winter.

* * *

I left Cape Hay, before the end of the Brünnich's Guillemots' breeding season but those few foggy weeks experience of the Arctic left me wanting more. Despite the problems of working in such remote areas I could see that, given slightly better living conditions, the potential for gaining a greater understanding of the biology of marine birds was enormous. Back in Britain I later heard that our habit of sleeping with a shotgun beside the cot had probably saved Larry's life at Cape Hay. One night soon after I had gone, an inquisitive or hungry polar bear decided to investigate the tent—an incident that Larry survived, but the bear did not.

CHAPTER 3

Nameless days

At breakfast time on August 24 [1899], we passed Princess Charlotte Monument and steered west through Lady Ann Strait. Here we had a heavy sea from the south-west with thick weather; though once when the curtain of fog was drawn aside, we caught a glimpse of Cobourg Island, and further up the strait, North Devon. Cobourg did not look very inviting as place of residence for human beings. All the way we went there were large flocks of guillemots splashing in the leads. We were sorry we had not been here in the nesting season, for what haul of eggs we should have then have got'.

This was Otto Sverdrup's (1904) account of his sighting of Coburg Island, lying at 76° North at the entrance to Jones Sound.

Eighty years later our approach was rather different, both in terms of transport and expectations. Our twin engined Otter started to descend onto an unpromising snowscape: a snow-covered beach, with a frozen sea on one side, and steep, glacier riven mountains on the other. The Otter landed on its skis in soft wet snow, and because he was anxious to be off, the pilot pushed all our gear and fuel out into a half-metre depth of slush, and disappeared. Ten minutes later when the drone of its engines had finally faded, we sat in the most intense silence I have ever experienced. We were on Coburg's Marina Peninsula under a cloudless sky and looking north over a dazzling expanse of endless ice. There was no wind, no sound from the frozen sea and there were no birds.

Coburg Island was originally thought to be a cape and named Cape Leopold by John Ross in 1818, after Prince Leopold of Saxe-Coburg of Germany (as was another seabird colony: Prince Leopold Island (Chapter 2)). The beautifully sculpted pinnacle lying off the end of Coburg's Marina Peninsula was named after Leopold's wife, Princess Charlotte. Once the cape was found to be an island in the 1850s it was called Leopold Island, and finally, to avoid confusion with Prince Leopold Island in Lancaster Sound, it became Coburg Island. The first reference to Coburg being an important seabird colony came from the diary of Charles Edward Smith on the whaling ship '*Diana*' written on 23 June 1866: 'We passed an immense "loomery" upon a large and rocky island opposite the entrance to Jones Sound' (Smith 1923; see also Nettleship, in press).

After moving our belongings to drier ground we waited with semi-frozen feet for a helicopter to appear and take us the few remaining kilometres to our camp. Many hours later, after the helicopter had slung all our belongings beneath its belly in a succession of tedious trips, we were finally left alone at what would be our home for the next 3 months. The camp location was spectacular, situated on a broad gravel beach, in a deeply curved bay, bounded at both ends by glaciers. The bay was frozen solid, but beyond it we could see the dark line of open water, and behind the camp scree-covered mountains rose

Marina Peninsula, Coburg Island with Princess Charlotte Monument at the far right.

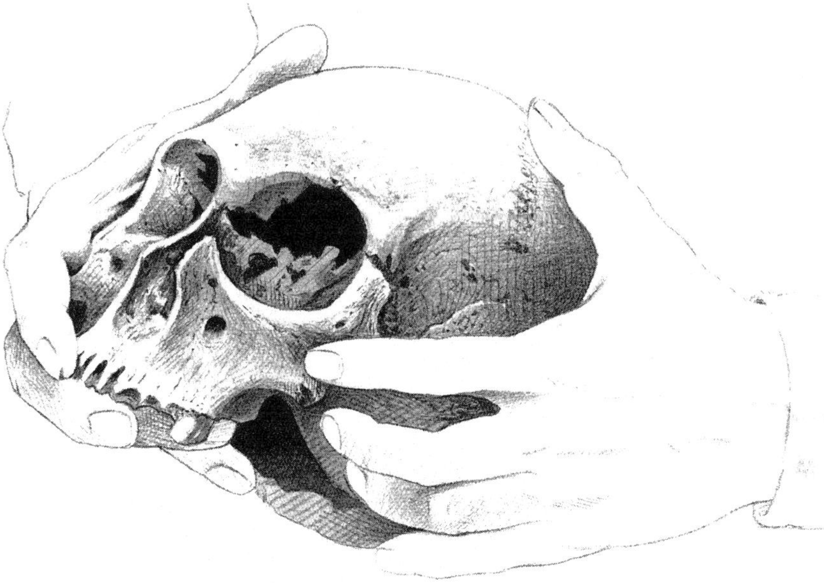

to 600 m. A stream running off the nearest glacier meandered past the camp and provided us with a constant supply of water. Three kilometres to the south was the Brünnich's Guillemot colony (Fig. 4).

In continual, bright sunshine we worked through the night to erect the Parcol and get the camp established. There were four of us, Don, Bill, Keith and myself, and Bruce, a young husky that we had rented from an Inuit in Resolute to serve as a polar bear early-warning system. Poor Bruce! He had spent his entire life chained up on a beach so his experience of life was incredibly limited, but then so was ours, and in retrospect, I realize we were all hopelessly prepared for what was in store. None of us were experienced with Zodiacs, yet our only access to the seabird colony was by boat. On our first morning after setting up camp, the wind coming off the glacier registered 60 km per hour on our anemometer, producing a spume-driven and distinctly uninviting sea. We decided to walk along the bay to see if there were any parts of the colony that could be reached on foot, since I realized that commuting to and from the colony by boat would be both dangerous and frustrating in terms of the days that we would inevitably lose. On both counts I was correct.

Our hike to the colony crossed one of the glaciers, and followed the beach round to the south. There were King Eiders, Common Eiders, and Longtailed Ducks (Oldsquaw) visible on the sea, Arctic Skuas (Jaegers) chasing Kittiwakes offshore, and the occasional Snow Bunting and Lapland Bunting on the beach. Half-way between our camp and the beginning of the colony we discovered an old Inuit camp, with the remains of two or three houses and stone caches. We had no idea of when the site had last been used, but guessed that it must be reasonably old. Tuskless Walrus skulls littered the ground around the camp and they had obviously been an significant part of the Inuit economy here. Seabirds must also have been important since the gravel around the site was composed almost entirely of Brünnich's Guillemot bones, representing countless numbers of birds, presumably killed and eaten over several successive seasons. Outside one of the small houses, I found the skull of an Inuit, bleached white, and broken across the orbits. I searched the area for the remaining facial section, but it was not until my next visit a week or so later that I found it and was able to fit the two parts neatly together. Fitting the two pieces together was like completing a circuit, and as I held the complete skull in my hands I tried to imagine what kind of life he or she had lived here.

The edge of the colony was disappointingly inaccessible. Looking upwards from the beach we could see large numbers of Brünnich's Guillemots and Kittiwakes distributed over the cliffs, but few of them could be reached by climbing, and none of them were suitable for observation. At this point the beach ended and cliffs fell steeply into the water, so we would not see any more of the colony until the sea calmed down and we got out in one of our Zodiacs. At the edge of the colony nearest the camp, we literally stumbled across the remains of a cairn, presumably made during the previous century. We searched it in the vain hope of finding the metal container, traditionally used to leave word of the expedition's past and anticipated movements, but found nothing. Interestingly, however, when I discussed this with David Nettleship many years later, he too had found a cairn the year before in the same area which *did* contain a metal cigarette box, but which was unfortunately empty.

* * *

The Brünnich's Guillemot colony at Coburg Island is one of only a handful in the high Arctic between Canada and Greenland, but all of them are extremely large and some contain as many as 300,000 pairs of birds. Further south, in boreal or temperate waters, the colonies of guillemots, and other seabirds, are more numerous but tend to contain fewer birds. The reason for the small number of very large colonies in the high Arctic (Fig. 3: see page 31) may be associated with the difficulties birds have in finding food in seas where ice cover and weather conditions are so erratic. The advantage of breeding in very large aggregations is that there will be an almost constant flow of traffic between the feeding areas and the colony, enabling the birds to keep track of the location of food as it shifts from day to day (Gaston and Nettleship 1981).

One problem of having large numbers of individuals distributed in just a few colonies is that they are incredibly vulnerable to man's industrial activities. A single oil spill in the vicinity of a high Arctic seabird colony during the breeding season could eliminate a substantial proportion of the total population. Because the Lancaster Sound region was thought to contain significant reserves of oil and gas, biologists were concerned about the effect that the exploitation of these could have on the ecology of the region. Our studies on Coburg were simply part of a massive ecological effort to document and investigate the many different, but inter-related, aspects of the wildlife of

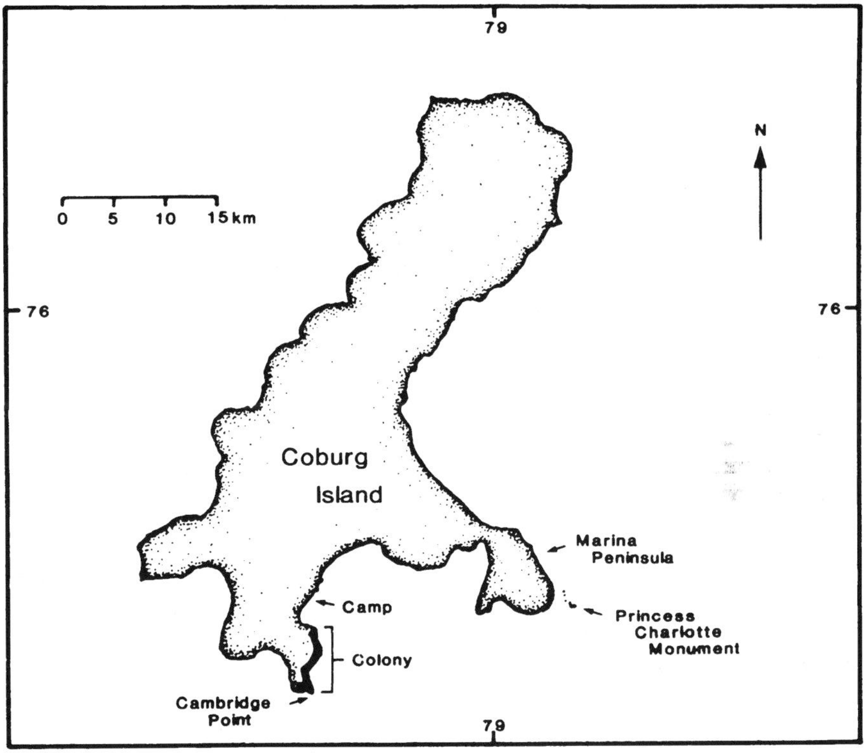

Fig. 4 *Map of Coburg Island showing locations mentioned in the text.*

this unique ecosystem. Regrettably, much of the money for such research came from the very companies that wanted to exploit the mineral reserves.

Our objective in this season was to compare the biology of two populations of Brünnich's Guillemots breeding at different colonies: one at Coburg and the other near Cape Hay on Bylot Island. I had been given the choice of returning to Cape Hay or being based on Coburg: on the basis of my experience the previous year I plumped for Coburg. At neither of these colonies had the biology of the birds previously been examined in detail and my role was to organize the scientific aspect of the studies, working from Coburg but keeping in regular radio contact with the team at Cape Hay.

* * *

On our second day the glacier wind ceased as suddenly as it had begun, and we carried a Zodiac and motors across the 500 m stretch of ice to the nearest point of land adjacent to open water. Once we were on the sea the full extent of Coburg's magnificent seabird cliffs become apparent. Unlike Cape Hay with

Part of the seabird colony at Coburg showing the remarkable density of Brünnich's Guillemots. Some Kittiwakes are also visible.

its 90° angles and muted colours, Coburg was a riot of form and colour. The 300 m high cliffs descended to the sea in variety of angles, intersected by an irregular series of canyons and gullies. High above the multitudes of birds, huge tracts of *Caloplaca* lichen gleamed like gold in the sun against the dark rock, and blue sky, contrasting nicely with the deep green of luxuriant stands of grass. Unbelievably, in the sheltered, south facing gullies, there was grass growing! At Cape Hay there was little more than mosses and lichens anywhere, but here on Coburg the plant life was patchily spectacular.

The colony of Brünnich's Guillemots and Kittiwakes extended along 6 km of cliffs (Fig. 4), with a short break in the middle, where the cliffs had disintegrated into scree slopes. The southern end of the colony was marked by a small, conical headland, a few hundred feet high, referred to as Cambridge Point. We landed here, tied up the boat, and I climbed as far up as I could, looking for eggs to see whether the Brünnich's Guillemot's breeding season had yet started. To my relief I found only a single egg and was pleased that we had not yet missed very much, but also that the breeding season was about to happen. As I climbed back down towards the Zodiac, on a rocky shelf beside the sea I found the bloody remains of a walrus calf. It had almost certainly been killed by a bear, and it was a poignant reminder of our vulnerability.

Returning back towards camp, we found a small, low-lying islet, joined to the main island by a 5 m stretch of ice. From the islet there were large numbers of birds close by, and I was beginning to think this would make a good study site. All at once there was the sound of whales exhaling close by and as we turned, a pod of 16 Beluga passed a few metres offshore, their glistening white bodies quite visible in the clear water. We never saw them quite as close again.

Adjacent to the islet on the main part of the island, the grass-covered cliffs reached down to the sea in a series of gentle steps. I climbed up about 60 m, and found the almost perfect study site on the opposite cliff wall. Below it, closer to the sea was another potential study site. We discussed the scientific and logistic merits of the three sites. From an observational point of view they were all good, and there was a lot to be said for them being relatively close together, since this meant less commuting and also that we were close enough to remain in visual contact. This turned out to be fairly vital on a number of occasions during the subsequent weeks.

* * *

The camp was luxurious by the standards of Cape Hay in the previous year. Our living accommodation was a Parcol, a sectional building comprising a robust aluminium frame, covered with insulated blankets. Each section came in its own box, measuring 2–3 m long, which subsequently became the floor. Overall, our Parcol was about 6 m long, and gave us plenty of room for working, cooking and sleeping. We had a four-burner gas stove, a refrigerator, deep freeze, an oil stove for heating the Parcol, tables and chairs, and proper beds. This might seem like decadence but it ensured that we were as efficient and productive as it was possible to be under Arctic circumstances. At our sister camp at Cape Hay, Erick, Gordon and Bruce were to make identical observations to our own. The two camps were connected by the invisible thread of radio waves and we spoke to them every day in order to co-ordinate our activities and to find out how they were getting on.

The camp on Coburg turned out to be one of the most harmonious and enjoyable that I was ever involved in, and I think this was due to the personalities involved. Don was a tall, fair-haired Canadian, an experienced rock climber, a keen wildlife biologist and someone for whom nothing was too much trouble. He was quietly confident and a meticulous observer. Keith was a bright young zoology undergraduate from Sheffield, full of boyish enthusiasm, a keen birder, and an excellent observer. He had the sort of inquiring mind which made me double-check my reasons for everything (scientific) we did. His Sheffield accent was such that I often had to act as interpreter for him with Canadians. Bill was a 16-year-old with no interest in the Arctic nor its wildlife and he had been sent along by a relative, in the hope that we could make a man of him. Despite his age Bill was built like an ox, and had

Aerial view of Cambridge Point on Coburg Island, the southern limit of the Brünnich's Guillemot colony. The colony is marked by the pale areas (of droppings) closest to the water. (Photo: D. N. Nettleship).

The Brünnich's Guillemot colony at Coburg. The author's hide was located at the top of the ridge running up from the right hand side.

undoubtedly made a man of himself many times previously in his native Toronto. He confided in the others and managed to shock them by telling them what his 'hobbies' at home were.

I decided to let Bill and Keith work together, in the innocent belief that being the two youngest they would have plenty in common and form a team. It did not quite work like that—largely, I think because their backgrounds were so different. Although Bill was still a teenager he had much more experience of life than Keith, and he was therefore much more independent. He could make bread and cook a three-course dinner without batting an eye-lid. Keith on the other hand was still living at home where his mum and four sisters had made sure he never wanted for anything and he barely knew how to make toast, let alone bread.

One morning shortly after we arrived, Bill was on breakfast duty, and with the culinary finesse of a chef he served each of us with a stack of perfect pancakes drenched in maple syrup. This was a new sensation for Keith and he decided to do the same the next day when it was his turn to cook. American pancakes are rather different from the things we call pancakes in Britain, and as I knew all too well, they required gentle coaxing in the pan. Needless to say we never got our pancakes, and Keith cleaned the batter from the Parcol walls when we came back from the colony later in the day.

I suppose Bill's prowess in the kitchen posed a kind of threat to Keith's position in the subtle camp hierarchy, but like so many chickens turned into a yard, they soon sorted out their differences, with Keith achieving the position he wanted. They both had a passion for cards, but lacking Keith's experience in this sphere, Bill was rarely able to win. This re-established Keith's dominance, but Bill's determination to improve meant that they played a great deal during their spare time and ended up as friends.

* * *

Keith and Bill's study site was the islet. Mine was a high ridge overlooking their site on one side, and Don's on the other. My position thus allowed me to keep in visual contact with everyone. We started off with three plywood hides, but as a result of various misadventures, only one of them ever ended up anywhere near the Brünnich's Guillemots. Fortunately, the weather was good for much of the time that we were actually at the colony and the hides were barely necessary. I had a wooden hide, Don improvised with some canvas and Bill and Keith sat in the open. Their's was an exceptionally sheltered position, and there were several occasions when I looked down to the islet to see Keith busy working and Bill laid out in his underwear, sunbathing! I felt sorry for Bill. He had been sent to Coburg under sufferance, he had no interest in what we were doing, but he remained cheerful throughout the summer. Because of his size and strength he was extremely good at carrying round the 40 hp outboard motors, and enjoyed driving the boat at high speed among the numerous ice pans.

We adopted a regular diurnal routine, working from 09.00 to 17.00 h on the cliffs, and going to bed at a reasonable time, despite the constant daylight.

In part our routine was dictated by our radio schedules with Resolute, and our evening talk to the others at Cape Hay. Part of our duty was to provide weather information to Polar Continental Shelf at Resolute twice each day to assist in the global weather forecasting process. Each camp in the Canadian Arctic did this and we were able to listen in on their reports. As in the previous year it became clear, so to speak, that Cape Hay was constantly swathed in fog, while much of the rest of the Arctic remained bright and sunny.

A regular problem in all camps, at least at the outset, is the toilet. We had been provided with copious sheets of plywood and two-by-fours, to construct a 'biffy', as it was euphemistically called. Don and Bill had spent the best part of day building it as soon as we arrived, but the glacial winds made sure it disintegrated a few day later. We then placed the 'can' in a sheltered position near the front of the Parcol and to ensure privacy, someone only had to say they were going 'seal-watching'. Indeed, the new position of the loo was perfect for bird and mammal watching. With binoculars one could look beyond the ice in the bay to see migrating seals, Belugas and the occasional Walrus. However, there were so many occasions when someone on the loo saw something exciting and called the others out to have a look that after a few weeks none of us had any inhibitions left, and no-one bothered to announce their intentions as they went out to the can.

There was a down side, of course. The toilet contents accumulated in a large plastic bag, which every so often had to be buried. Everyone volunteered to dig the necessary pit, which was several hundred yards away from the camp, but no-one offered to empty the loo. As camp leader, this delightful job fell to me. The real problem was that it was only the bag's contents that we wanted in the pit, and not the bag itself. This required some extremely delicate manoeuvring, especially when there was a stiff breeze blowing.

The first 2 weeks on Coburg were idyllic. The sun shone all day and all night, and the sea was like glass and free of ice, so commuting between camp and the colony was straightforward. The outboards behaved themselves, and we were collecting plenty of information from the birds. Naively, we expected this to last, but of course it did not.

* * *

Spring comes in a rush in the Arctic and it was in full swing when we arrived. The sheltered gullies were full of colour, with dandelions, stitchwort and cinquefoil all in bloom. Such locations offer the best of all possible worlds for plants: the snow lies late in the gullies and protects them from low winter temperatures, but in the spring it also ensures a supply of precious water, and the south facing slopes catch the best of the sunshine. On our beach near the camp there were Arctic poppies, lousewort and horizontal scrub willows. On the ornithological front Coburg Island fulfilled all the biogeographical rules which meant that our bird list was abysmal. Apart from the birds mentioned earlier the only highlights were a pair of Long-Tailed Skuas (Jaegers), a single Arctic Tern and a Thayer's Gull.

One cool morning at the Brünnich's Guillemot colony, when snow flurries alternated with bright sunshine, a male Snow Bunting sang his cheerful, tinkling song from the top of my hide. As he flew off I opened the door to watch him in his crisp black and white plumage. It was the first time I had heard a Snow Bunting sing and it brought to life an account of Greely's ill-fated expedition, not too far north of Coburg, on Ellesmere Island.

During the first International Polar Year in 1882–1883, eleven countries established scientific stations in the Arctic specifically to collect meteorological and magnetic information. America's contribution was to run the most northerly station, at Fort Conger in Lady Franklin Bay on the north-eastern tip of Ellesmere Island. Adolphus Greely, aged 38, and his party of 24 men were dropped off at Fort Conger on 11 August 1881 on the understanding that they would be re-supplied the following year. The relief vessel, however, failed to break through the barrier of sea-ice 320 km to the south and although the party was not short of supplies, the failure of the ship to appear demoralized the men. Despite this set-back Greely's men clocked up three new Arctic records: farthest north, east and west during this period. A year later, in August 1883, there was still no sign of a rescue vessel, so Greely and his men set off for the south in the hope of meeting the boat en route, and where he anticipated finding some pre-arranged caches. They saw no boat, and the food-caches they found were totally inadequate. The men began their third winter in the Arctic in poor shape and dismal spirits. Greely's (1886) description of that winter makes pitiful reading: starvation, frost bite, execution for theft, and cannibalism.

> On Easter Sunday we heard on our roof a snow-bird chirping loudly—the first harbinger of spring. All noise stopped as by magic, and no word was said until the little bird passed. His coming on the Sabbath morn was thought a good omen, and did much to cheer us through the day (Greely 1886).

By June 1884 Greely and the six other survivors were reduced to eating lichens and gnawing oilskins. Rescue finally came in late June, and six of the seven survived to tell (selected parts of) their desperate tale.

* * *

Every day for 3 weeks we walked across the ice in our bay to the tip of the north colony where we kept our Zodiac and motors. Every evening after returning from the colony we deflated the boat, and every morning we pumped it up again. We performed this wearisome exercise, not for pleasure, but to minimize the risk of an inquisitive bear puncturing the boat, as they were known to do. We looked forward to the day when the ice in our bay would disappear and allow us to drive right up to the front door of the Parcol. Eventually that day came, but not in the way we expected. We were confined to camp because of unbelievably strong winds which drove the needle on our hand-held anemometer to the limit of its scale and kept it there (over 70 km per hour). With all our camp chores completed, Keith and I had walked north of the camp to investigate the glacier where it met the sea. On looking out across the bay we were surprised to see that one or two leads had appeared in the ice. Less than 15 minutes later the entire bay was ice-free: huge sheets of ice,

Getting to our study plots by inflatable through the ice. Cambridge Point in the distance. (Photo: D. Reid).

measuring several hundred metres across, broke loose and were blown out of sight towards the Greenland coast with alarming rapidity. As we watched, we reflected on how many times we had walked across that ice and never considered the possibility of it disintegrating so rapidly. I also wondered whether we would have noticed in time if we had been on it when it started to break up!

* * *

The view that Don and I got from our respective hides was confined to the birds on the cliff-face in front of us, Brünnich's Guillemots and Kittiwakes in my case, and Brünnich's Guillemots alone in Don's. Without a hide, Bill and Keith's location on the islet gave them much greater all-round vision. They could see the huge scree slope that separated the two halves of the colony, the hump of Cambridge Point and beyond that, the distant glaciers of Devon Island, 50 km to the south across Lady Ann Strait. They also had a fine view of the full extent of the colourful and architecturally complex cliffs towering above their study plots. It was not entirely perfect, however, and one of the slight disadvantages peculiar to Bill and Keith's islet was that it lay below the flight-line of thousands of Brünnich's Guillemots. So much so that they found it almost impossible to eat their packed lunch without also ingesting at least one mouthful of Brünnich's Guillemot droppings each day. When they first complained about this I felt they must be exaggerating, but on joining them on their tiny rock one day, I found it to be true. I was mildly surprised because I had assumed that the birds leaving the cliffs simply flew out to sea in a straight line, rather than having particular routes, which they obviously did.

It was hardly surprising that Keith and Bill produced the best bird and mammal records for our Coburg species list. This was partly because Bill spent more time looking anywhere except his study plot, and partly because having seen something he could exploit Keith's twitching ability to tell him what it was. On one occasion Bill spotted something he was not expecting. As he scanned the water towards Cambridge Point, he saw a polar bear swimming across the bay. Both he and Keith were extremely excited, since this was the first bear they had seen. They shouted to me, but their cries were lost among the voices of tens of thousands of Brünnich's Guillemots and I remained oblivious in my hide. The bear swam slowly and steadily in a straight line adjacent to their islet, heading towards Princes Charlotte Monument, the most easterly point of Coburg. Keith and Bill watched through their binoculars and telescopes, revelling in their first sighting of this magnificent predator. Apparently unaware of their presence, the bear had a sudden change of plan a hundred or so meters away from the islet and made a 90° turn towards it. Keith and Bill were ecstatic and started to take photographs. They were still taking photographs when the bear pulled itself out of the water beside the Zodiac. It was not until this instant that they simultaneously realized there might be a problem. The Arctic is full of stories, most of them true, about bears attacking and killing people. Indeed, we had all practised using the shot-gun, precisely for incidents like this. But of course, we had long since given up carrying the guns in the boat each day. We had seen no sign of bears, and the constant splashing of seawater caused the barrels to rust overnight.

When the bear was just 5 m away their ecstasy turned to anguish, as they abandoned their cameras and started to scream and shout at it. There was nowhere to run to; the islet measured no more than 20 m by 10 m, and if the bear decided to go for them, one or other of them would be sure to be caught. They continued to scream and wave; the bear lowered its head, and swaying from side to side, it hesitated. Then, miraculously, it turned and got back in the water. At this stage Don and I were still oblivious of this drama. Once the bear swam away and was several metres from the islet, Bill and Keith jumped into the Zodiac and started the engine. This low amplitude sound broke through the raucous cries of the Brünnich's Guillemots, and I jumped up, wondering why anyone was in the boat so far ahead of 'going home time'. I looked down through my binoculars, but could see nothing untoward, except that Bill and Keith were pointing excitedly at the water in front of them. There were many small pieces of ice about, and I still could not see the bear, until they zoomed right up to it in the boat. I immediately climbed down to the foot of the cliff to meet them, and Don, where they described their close escape. Oscar Wilde once said that 'only the careless have adventures', but there really was not much that was careless about this particular encounter. The only thing we should have done was to carry the guns with us, which we did from then on. We could only assume that this was either a female bear (females are generally more wary than males), or one that had had a bad experience with people previously. If it had been a young male, there might not have been a story.

The bear continued to swim to the northeast, and the last we saw of it was one and a half hours later, 8–10 km away still heading out to sea. Luckily, that was our one and only encounter with a polar bear on Coburg; our sister crew, at Cape Hay was less fortunate.

*   *   *

Apart from our bear there were few predators on Coburg. There were arctic foxes but we saw them only infrequently. Once a fox climbed down the steep slope to Keith's study plot, grabbed and killed an adult Brünnich's Guillemot and climbed back up over the cliffs again with its prize. There must also have been a fox den on the buttress where my hide was situated although I never saw an adult there. One afternoon as I climbed down to the boat I was surprised to hear a soft mewing call coming from somewhere near my feet. The sound had a ventriloquial quality and it took me several minutes to locate the tiny fox cub staggering around the cliffs. Whether this was his first outing, or whether I had disturbed the adults in the process of moving him, I could not tell.

We did not see a single bird of prey during our time on Coburg. Instead, the raptorial niche was filled by those sullen, confident opportunists, the Glaucous Gulls. Mafia-like, the gulls had divided up the Brünnich's Guillemot colony so that each pair had a territory from which to extort a living. Near each gull nest there was a rocky promontory from which the gulls kept watch. Just as gulls elsewhere, they were extremely aggressive, and we usually had to carry an oar from the Zodiac with us to keep them at bay when we checked their nests. Each pair of gulls seemed to specialize in one particular type of prey. The pair at Keith's study plot were Kittiwake specialists, and their strategy was to fly in to the colony, grab a Kittiwake off its nest and drag it down on to the sea and drown it by holding its head under the water. Only then, the second member of the gull pair would alight on the water and the two of them would reduce their unfortunate victim to bite-size pieces.

Other pairs of gulls seemed to subsist without ever killing anything themselves, but scavenged the corpses of other birds killed by rock-falls. In certain parts of the cliffs there was an almost constant shower of rocks, ranging in size from peppercorns to footballs. As you might expect, Brünnich's Guillemots and Kittiwakes avoided breeding in areas with regular rock-showers, but there were still victims, and occasionally huge sections of cliff would detach themselves for no obvious reason, killing dozens of birds.

Several pairs of gulls specialized in Brünnich's Guillemot eggs, and their nest sites were littered with empty egg shells. The majority of the shells were unbroken because the gulls typically removed and ate the egg's contents through a relatively small hole. Since we could measure these shells I was interested to see how their size compared with a sample of unpredated eggs from the cliffs. Risking attack, we made regular visits to these particular gull nests and I asked Erick's team to make similar measurements at Cape Hay. There was an interesting difference between our results: at Coburg the eggs taken by the gulls were much smaller, on average, than those we measured on the cliffs. The most likely reason for this result was that young or poor-quality Brünnich's Guillemots laid small eggs and bred in places where the gulls could steal their eggs. At Cape Hay, however, there was no difference in the size of predated and unpredated eggs. Ice-falls were frequent at Cape Hay and when they occurred, the birds simply abandoned their breeding ledge leaving their eggs unattended. The gulls would then alight on the breeding ledges and take whichever eggs they wanted.

* * *

One of our tasks at Coburg was to collect information that would allow us to interpret photographs taken of the entire colony. The most basic information necessary for the conservation of any species is an assessment of its population size. One way of estimating the size of large seabird colonies is to take a series of overlapping photographs on a large format camera, to make large black-and-white prints and then for some unfortunate soul to count the birds. Since the structure of Coburg's cliffs meant that very little of the colony was visible from the island itself, the only way to tackle the Brünnich's Guillemot census was to fly along the length of the colony and photograph it all. Even assuming you can do this successfully, and can later count birds (usually mere dots) on the photographs, it is essential to know what the final numbers mean. Brünnich's Guillemots make no nest, so one cannot count nests to estimate the number of breeding pairs, and the numbers of birds on the ledges at any one time is not the same as the number of breeding pairs. To make sense of the photos one has to do two things. First, set up study plots, where detailed daily observations over the egg-laying period provide an accurate count of the number of pairs breeding on particular areas. By later counting the total number of individual birds on those same areas you can then derive a correction factor, to convert total numbers of birds to breeding pairs. However, it is not even as straightforward as this since the number of birds at the colony varies over the 24 hour cycle. At colonies like Skomer, in temperate regions all the birds that are not incubating an egg or brooding a chick leave the colony at dusk and roost on the sea, returning at first light the next day. There was no dusk or darkness on Coburg so the second thing we had to do was to see whether the number of birds at the colony varied over a complete 24 hour cycle. With this information, it does not matter what time of day the census photographs were taken, because you can correct for the time of day to convert number of individuals to breeding pairs.

Our study plots were established and we knew how many pairs each contained. All that remained was for us to do was to determine the diurnal fluctuations in numbers. We decided that we should use just two of three study plots, enabling us to work in pairs, and that we should adopt 12-hour shifts, counting the birds every hour. Keith and Bill would drop Don and I at the colony at noon, and then at midnight they would return to change places with us, continuing the counts for the second 12 hours.

We were duly deposited at the colony. It was Friday 13 July. I am not superstitious, but on this occasion perhaps I should have been. The weather had been reasonable when we landed, but within an hour of our starting the counts it started to snow, and the wind picked up. It snowed horizontally, and no sooner had it started when all the off-duty birds began to leave the cliffs. Mentally I groaned, since I knew what this meant. Remarkable as it may seem, Brünnich's Guillemots appear to be able anticipate bad weather, probably by being sensitive to changes in barometric pressure. For some reason they would rather be on the sea than at their breeding colony during inclement conditions. I realized that as soon as the birds who were not incubating left the colony, they probably knew more about what was in store than we did. As a result, within an hour all that was left on the ledges were birds with eggs, and the numbers in my note book remained identical hour after hour, and were completely useless in terms of what we were trying to do.

My hide provided some protection from the weather, but because it had a large hole (a 'window') cut in the front of it, the snow started to blow directly on top of me. I managed to fix my rucksack over most of the window, and to stuff the remaining gaps with the bags we used for putting birds in. Every hour, I dismantled my barricade, and made another identical count of the birds on the cliff opposite. The birds seemed resigned to the weather. They hunched themselves up, closed their eyes, and sat tight. As soon as the snow started the temperature dropped and the combination of boredom, irritation and sitting still left me feeling uncomfortably cold. In contrast, I do not believe the birds even noticed the temperature. After all, this was summer (I kept telling myself), and they have to endure far lower temperatures than this at other times. Experiments with Brünnich's Guillemots in captivity have shown that they are able to withstand temperatures as low as −55°C for 18 hours with no drop in body heat, due mainly to a high metabolic rate rather than particularly good insulation (Johnson and West 1975).

The hours ticked by, but eventually as midnight approached my morale increased in the anticipation of soon being picked up by the others. I took my rucksack from the window, packed my belongings, and climbed out of the hide into an unfamiliar, white world. I made my way down the slippery grass and snow-covered slope to a point where I could see Don just emerging from his makeshift canvas hide. Although we could see each other, we were dependent upon the boat to get to each other's study plot. Less than 50 m separated his hide from the point I was now at, but we could not make out what either of us was shouting. Ready and anxious to get back to camp we both stood there, stamping our feet in chilly expectation. As the minutes dragged past midnight I became less and less optimistic. It was still snowing hard, but the sea was quite reasonable enough for the boat to get to the colony. After 30 minutes of standing around, losing valuable body heat, I signalled to Don that we should go back and wait in our hides—at least we would be marginally warmer there.

With a glimmer of foresight we had left a small cache of food and three sleeping bags at the foot of the cliff near my hide. If we had had better foresight we would have made sure there was a cache at each of the three observation points. With a mixture of relief and guilt, because I knew Don could not get one, I took a sleeping bag from the cache and went back up to the hide. A great deal of snow had blown in during the hour that I had been away, and I scraped out as much as possible to prevent the sleeping bag from getting wet, then re-rigged my rucksack over the window. I got into the sleeping bag, positioned myself diagonally in my four foot cube and tried to get to sleep. After sleeping fitfully I woke at 08.00 the next morning. It was still snowing. I climbed down the cliff to where I could see Don. He climbed out of his hide and we shouted to each other, but there was a tremendous swell with surf crashing on to the rocks below us so it was impossible for us to communicate verbally. Don gave me a thumbs up sign to indicate that he was okay, and he went back into his hide. By now I was starting to get anxious, wondering what could have happened to Keith and Bill to prevent them from coming for us for so long, and I was worried about whether Don was keeping warm.

During the night, my makeshift cover for the window had moved in the wind and my sleeping bag now had large wet patches on it where the snow had settled. I therefore took a second sleeping bag from the cache and again went back to the hide. I had long since given up counting the birds and I set about trying to fix the window, and used one of the two shirts I was wearing to keep the snow out. As I did this I glanced over to my study plot and could see that many of the birds were almost completely buried in snow.

It was a long, long day. With the window effectively sealed off at last, it was almost totally dark inside the hide, and far too cold and windy to face opening the door in the lee of the wind to get some light. I was warm inside the sleeping bag and dozed intermittently, mentally running through the possible reasons, positive and negative, for Keith and Bill not being there. In my shallow sleep I

The hide which became a home. This photo was taken shortly before we climbed out over the cliffs.

had a succession of fast and tumultuous dreams. Suddenly my head was pounding and with adrenaline streaming through my blood vessels, I dragged myself out of my dreams into wakefulness, but the pounding continued. In the disorientated moments between sleep and waking, I could hear Don calling, outside the hide, and thinking that the others had finally arrived, I pushed open the hide door. It was just Don, and I felt a mixture of elation and disappointment. I was pleased to see him, but discouraged by the absence of the others. Incredibly, Don had managed to climb from his study site, to mine, using a 2 m piece of rope he happened to have in his rucksack. Earlier in the season we had looked at this section of cliff and decided that it would be impossible for anyone to make that climb. The combination of cold and hunger had forced him to try, and despite being an experienced climber it had still taken him over an hour to scale a relatively small, but vertical section of cliff. By the time he arrived it was 06.00 h on our third day.

Don collected the third sleeping bag, a small primus stove and some of the food from the cache, and squeezed into the hide. I knew from the accounts of polar exploration that obtaining drinking water from snow or ice was tricky; now we were able to experience this first-hand. We opened a can of pineapple, ate the fruit and packed the tin with snow to make ourselves a much needed drink. Our tiny primus roared away for 30 minutes to give us half a mug of water each. We had a jar of honey, and made ourselves a hot, sweet drink. Just being together boosted our spirits enormously, and we sat, warm but cramped in our sleeping bags. We chatted and exchanged ideas regarding the others.

We slept through the night, and by the following morning the snow had stopped, the wind had died and the sea was relatively calm and free of ice. We got out of the hide to stretch, and decided that if Keith and Bill did not come under these conditions then something serious had certainly happened. We waited, but the welcome sound of the outboard never came. The birds on the ledges continued to sit on their eggs. As the snow started to melt, the breeding ledges became a quagmire of pink sludge, and the birds started to take on a dishevelled appearance once their feathers were matted and wet, and bespattered with droppings. Things were getting serious for them too; I knew from my previous observations that 4 days is about how long a Brünnich's Guillemot can last without feeding. Birds which lose their partner generally give up caring for their egg or chick after 4 days of incubating alone. The entire colony was effectively in that position now, and there was no sign of anyone coming to relieve them either. However, to be honest, by this stage I was more worried about ourselves than the birds.

As we waited through the morning the weather closed in again, and with it our optimism disappeared. We discussed the possibility of climbing out, up the towering cliffs above us, but it did not even seem like an option to me. By midday there seemed to be no alternative but to go and see how feasible the climb was. Before setting off we ate a 'big' lunch, sharing a tin of corned beef and a tin of apricots, followed by hot honey solution from the apricot tin. We then left a note in the hide detailing the direction in which we had headed. We took nothing unnecessary with us, leaving our once-precious note books in the hide, along with our binoculars and other 'useless' belongings.

Outside the hide, the snow-covered grass extended for a further 50 m up towards the first cliff face. As we scrambled up to it we wondered whether that might be as far as we could get, but to our relief, there was an easy way round leading to a long grassy gulley. It was very slippery, but we climbed slowly and carefully, eventually reaching a pinnacle of rock. Again we were able to climb up and over it without too much difficulty. There was a succession of pinnacles, each one connected by a sloping knife-edge of gravel. Each time we approached a pinnacle we wondered whether it might force us to turn back, but the wet, knee-deep snow enabled us, with one foot on either side of the knife-edge, to ascend. Without the snow, I doubt whether it would have been possible. Fortunately, all the pinnacles were interconnected in this way, enabling us to get further and further up the cliffs and away from the hide. Every so often we stopped for breath, and turned to look back the way we had come and towards the south end of the colony: from our novel angle the view was magnificent.

After one and a half hours of climbing, we reached the top, only to find ourselves surrounded by fog, in a flat, white, featureless landscape. Luckily, Don and Keith had walked and climbed in this general area a week or so earlier, so Don was reasonably confident about the direction we should take. It was easy walking now, and we continued on fairly level terrain for the next 40 minutes. Because of the fog it was difficult to know where we were, but we were aiming for the north end of the colony, albeit at a considerably greater altitude than we had previously experienced.

Quite unexpectedly there was a momentary break in the fog, and we were able to make out a huge white area far below us. As we struggled to interpret what we saw, it suddenly all fell into place: the white area was the glacier by our camp. Before we could see anything else the fog closed in again, but we knew then that we had made it. Our next task was to find a way down from our mountain top, to the bay below. Retracing our steps we found a snow field sloping steeply down towards the glacier. As we discussed the best way down, there was another, longer break in the fog, enabling us to see the Parcol, and the entire bay. It was completely choked with ice, so it was little wonder that Bill and Keith had been unable to reach us.

Excited, we virtually slid the entire way down the snow field to the shoreline. As we crossed the glacier, once more on familiar territory, we could see Keith at the stream collecting water. We shouted to him and Bill appeared from the Parcol, and they both ran over to greet us. After sighs of relief all round we compared our respective stories.

Although we were tired, we were in reasonable shape. Don, who had been wearing wellington boots, had frozen feet, but luckily suffered no long term damage. On reaching camp we suddenly realized how thirsty we were, and our first response was to drink several litres of water each. It was rather perturbing to realize how dehydrated we had become without recognizing the fact. Then, since it was supper time, Don and I sat down to a huge meal that Bill and Keith (probably mainly Bill) had prepared. As we did so, Keith got on the radio to tell the people at Resolute that we had made it back okay. Don and I were completely unaware that at Resolute, Polar Shelf had been waiting for a break in the weather to send a helicopter out for us. Now that we were back I was very glad they had not, since it would have put paid to our study!

* * *

The ice that had blown into the bay by the camp remained for a further 5 days. It also continued to rain and snow during that time, so we were unable to get back to our study plots at the colony. This was both frustrating and disappointing. I was anxious to know what had happened to the birds, and worried that a 5 day gap in our data would render our study worthless.

At Cape Hay in the previous year I had been impressed by the variation in the size of Brünnich's Guillemot eggs and by the difference between the size of eggs between Cape Hay and Prince Leopold Island. One of the objectives of our studies this year was to try to find out what effect egg size had, if any, on the survival prospects of the emerging chick. In humans, birth weight is closely associated with survival: very small and very large babies are much less likely to survive than those closer to average weight.

The size of egg an individual female lays is likely to be affected by a number of factors, including its body size (big birds may lay big eggs), and its condition (birds in good condition may lay relatively larger eggs). In addition,

I was interested in the possibility that Brünnich's Guillemots might adjust the size of egg they produce according to the time of season when they lay. Studies of several other bird species had shown that eggs laid early in the season tended to be larger than those laid later. I wondered if the same might be true for Brünnich's Guillemots, and if it was, what could explain it. A simple explanation might be that poor quality or immature birds tend to breed later in the season, lay smaller eggs and hence produce this effect. On the other hand, if I could persuade some early breeding birds to lay later than they would otherwise do and see what size eggs they produced, it might be possible to disentangle the effects of laying date and bird quality. Since Brünnich's Guillemots readily lay a replacement egg if their first is lost or taken, this seemed like a relatively simple experiment.

Quite close to my hide was a broad, sloping ledge with a hundred or so pairs of Brünnich's Guillemots breeding on it. When the first 20 pairs there had laid I climbed down onto the ledge and removed their eggs, taking care as I did so, to number each egg and the precise spot where each egg had been laid. I used a large magic marker pen to number the sites, and on the warm, dry day on which I did this it seemed fine. The numbers stood out clearly on the rock, and I assumed I would be able to return under similar conditions to measure the replacement eggs. As a precaution however, I also wrote numbers on the underside of large rocks adjacent to each site.

Back in camp we weighed and measured all the eggs, and then went about our other field work while we waited for the birds to re-lay. It usually takes Brünnich's Guillemots about 14 days to produce a replacement egg, but the 14th day coincided with Don and I being stuck at the colony and it was not until later that we were able to get back to check for replacement eggs.

On the 20th day the ice finally disappeared from the bay beside camp and although there was still a very heavy swell, we decided to see if it was possible to get to the colony. The boat journey was fine, but at the colony we were disappointed to find that a lot of ice had blown up against the cliffs, and it looked as though it was impossible to land. However, when we got there we found one part of the islet to be ice-free, so we landed there. I was able to get across to my study area because a large piece of ice had wedged itself into the watery gap between the islet and the main island. I climbed up to the experimental ledge and was pleased to see that all the birds had re-layed. As I crawled on to the ledge I looked anxiously to see if my magic marker numbers had survived the rain and snow, and to my relief found that they had. I started to measure the eggs, but after measuring only three the weather began to close in, and I knew that with the prevailing ice conditions the way they were it would be prudent to get back to camp. I was then faced with a dilemma: my original intention had been simply to measure the replacement eggs and leave them to hatch, but I knew that if I left the eggs now and if the inclement weather continued I might never get a chance to measure them. In addition, I was worried about the gaps that the weather had already punched in our data and feared that we might end up with no useful information for all our efforts. Twenty eggs out of a total of several hundred thousand in the entire colony seemed almost insignificant, and we knew from the rock-falls that we had witnessed that hundreds of eggs could be lost in an instant. Gritting my teeth, I decided simply to take the eggs so that I could measure them back in camp. It was a rapid decision, and probably is not the one I would make now.

I placed the eggs in a large cloth bird bag and started to climb back towards the boat. The rain had made the grassy slope very slippery and I was forced to carry the bag in my mouth so that I could use both hands for climbing. As I got down to sea level I could not see any of the others, and to my horror I found that the ice pan that I had used as a bridge to the islet had disappeared. There was a spectacular swell and further down the shore huge pieces of ice were churning and grinding into each other. As I dithered on the 'wrong' side of the channel, a piece of ice drifted in and I took a gamble and jumped on to it, in the hope that I could encourage it to move in the right direction. This was rather stupid since the amount of effort needed to move the ice was far greater than I could exert. After 15 frustrating minutes, still with the bag of eggs in my mouth, the ice pan did move somewhat closer to the islet and I managed to get ashore with nothing worse than wet feet.

The others were on the far side of the islet watching the ice war that was raging in the bay between the north and south colonies. Huge, flat pans of translucent blue ice jarred and butted each other. Occasionally, the sea would break over them, or a pan would savagely mount another, only to slip back into the water, reminding me of so many enormous, frenzied, randy beetles. The boat was pulled well up out of reach of the ice, but I knew we could not stay on the islet indefinitely. The bedlam in the water was such that the ice was shifting position all the time, so we decided that we would launch the boat and try to break through into open water at the first opportunity. After a few minutes our chance came and, still struggling with my precious cargo of eggs, we leapt into the boat and set off. Zodiacs have the big advantage of being fairly flexible and compressible, and this particular attribute was never more valuable. Outboard motors on the other hand have an impeccable sense of timing, and ours decided that this would be a good time not to start. We drifted helplessly amidst the churning ice, all the time frantically trying to pull the engine into life. The ice squeezed us into a narrow lead alongside an enormous, rather stable looking ice pan. I wondered whether we would not be safer getting ourselves and the boat out of the water and onto the ice pan, but our 'discussion' of the pros and cons of this was rather fraught, and took place at high volume and without many of the normal conversational conventions. The engine suddenly exploded into life and the ice made our decision for us. A good-sized gap appeared between ourselves and the open sea and we agreed to go for it. Don revved the engine to full power and we set off at high speed, but even as we did so I could see the gap starting to close. When we reached it the opening was little more than a metre across, and with huge pieces of ice squeezing us on each side, we popped through into open water, rather like a piece of wet soap being 'shot' from the hand. The gap between success and failure here had been a fine one, and the natural response to the former is always elation. We set off back to camp congratulating ourselves on our success, although I reserved some elation until we got back to camp and determined how many eggs had survived.

Remarkably only a couple of the eggs were broken, and when I compared each original egg with its replacement the match of colour and markings was dramatic. Also, as I had hoped, the replacement eggs were smaller than the first eggs, by about 6% on average (Fig. 5). This indicated that even good-quality birds reduced the size of their egg if they were forced to lay later than they would otherwise have done. The next stage was to try and discover why this was so.

I wanted to test the idea that there was a trade-off between the size of the egg and when it was laid. Our results showed that those birds which laid early, produced big eggs and were most successful in rearing chicks. Why then did not all birds lay as early as possible and produce big eggs? You might as well ask, why are we not all capable of running half marathons? The answer is that like us, Brünnich's Guillemots (and other organisms) differ in their quality and, in this particular example, not all birds are in good enough shape to lay large eggs early in the season. I was interested in the way that evolution had produced a solution to this particular problem.

Two factors seemed to be important in determining whether a chick fledged successfully or not: the time in the season when it was born (that is, when the egg from which it hatched was laid), and the size of egg it hatched from. Because the Arctic breeding season is short and environmental conditions rapidly deteriorate towards the end, the later a pair lay the less likely they are to be successful in rearing a chick. One reason that chicks which hatched from big eggs did better than those from smaller eggs was that the former contained a greater reserve of yolk, giving them a head start in their development.

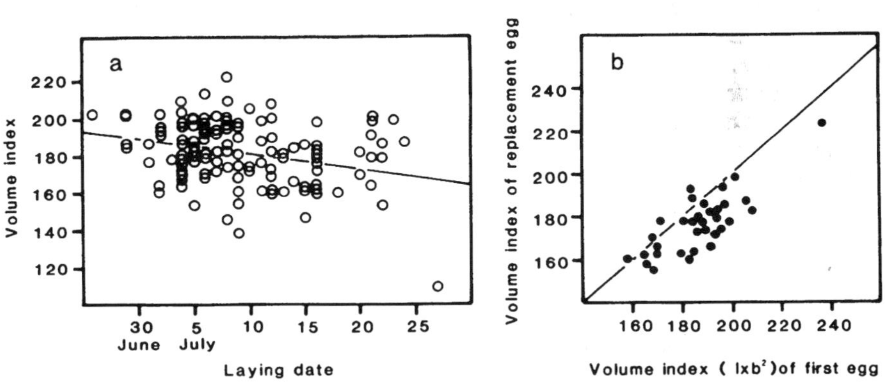

Fig. 5 *(a) Changes in the size (volume index:* $1 \times b^2$*) of Brünnich's Guillemot eggs laid on different dates at Coburg Island. Overall there is a significant decline in egg size through the season: later laid eggs are smaller than those laid earlier. (b) The relationship between the size of first and replacement eggs laid by the same female Brünnich's Guillemots at Coburg Island. The trend is significant indicating that birds which lay large first eggs also produce large replacements, but the fact that most points fall below the 1:1 line shows that replacement eggs are generally smaller than first eggs. This effect may contribute to the effect shown in (a). From Birkhead and Nettleship (1982, 1984).*

Our results also revealed that the timing of laying had a much greater effect on the likelihood of chick survival than did egg size, presumably because parent birds have the potential to compensate for a small yolk reserve in their chick by giving it more food once it hatches. Not surprisingly then, it was more important for birds to lay as early as possible, rather than to delay egg laying, in order to make a larger egg. We found therefore that the trade-off is a simple one: birds unable to lay at the very beginning of the season should lay as just soon as they were capable of producing an egg of viable size and this is exactly what they appeared to do, hence the decline in egg size through the season (Fig. 5).

The birds on our study plots, and presumably elsewhere in the colony, had weathered the storm, with no apparent cost. Perhaps the Brünnich's Guillemot's impassive manner has evolved precisely to get them through difficult times like this, when tenacity pays off. When we next checked the birds there were few losses, indicating that despite the miserable weather conditions, and my worst fears, the birds had managed perfectly well.

* * *

To our surprise there was large ship on the horizon. From our study plots we each examined it through a telescope, only to see a large military helicopter take off from its stern. Fascinated, we watched as the helicopter flew straight towards us. It flew past us and towards our camp. A few minutes later it returned, now flying close, much too close, beside the colony. Brünnich's Guillemots started to leave the cliffs in great showers, and the helicopter, obviously pleased with this effect, flew in closer, causing immense numbers of birds to cascade from the cliffs. The air was filled with the stench of exhaust fumes and with the throbbing sound of the machine's huge rotors. At least one bird was killed by the helicopter which was so near we could see the faces of the men inside. We gesticulated wildly, making sure our sentiments were unambiguous, and the helicopter eventually buzzed back to its mother ship. It took several hours for our anger to subside; their violation was typical of the military's arrogant disregard for life.

* * *

Commuting to and from the colony each day made us preternaturally observant of prevailing weather conditions. At home, working inside, one uses a mental scale of, say 1 to 5 to rank the weather. Here, where so much depended on it, our scale stretched over a much greater range of values and while we were at the colony we monitored every nuance of the meteorological conditions. There were aesthetic benefits: never before had I been so conscious of cloud formations and the effect that even small topographic features, such as the Marina Peninsula, had on micro-weather systems. On several occasions the direction of the wind would change while we were at work, bringing expanses of sea-ice drifting in towards the colony. We would then beat a hasty retreat back to camp to avoid being trapped. Sometimes the sea-ice brought us walrus. For some reason, the pans that carried walrus rarely came within a kilometre of the cliffs, but through binoculars or a telescope we could see their fat formless brown bodies piled up against each other on slabs of soiled ice.

Only twice did we see walrus close inshore. The first was when Don and I were precariously perched on a cliff weighing Brünnich's Guillemot chicks at the north end of the colony. A female walrus and her calf swam directly below us in the clear water. They both surfaced together and oblivious of our presence, cruised gently along the rocky shore investigating all the small bays. On the second occasion we were at the study plots and were distracted by a loud, rhythmical grinding noise. It sounded almost as though a helicopter would appear at any minute, but it never did. The noise continued and was louder than ever as we climbed into the Zodiac. As we pulled away from the islet the head of an enormous bull walrus emerged a few metres behind in our wake. The noise was his threat display, and we considered ourselves lucky that he had come up only after we had passed over him. Walrus use their massive heads to break through ice and have a reputation for overturning boats.

Boating to the colony each day certainly had its disadvantages, but it also had a few benefits, and one was being able to witness at first hand the remarkable productivity of polar seas. All through the perpetual winter darkness, the bodies of dead organisms, small and large, accumulate in the sea. But trapped beneath the ice and starved of light, the phytoplankton, the first links in the marine food chain, are unable to exploit this rich source of nutrients. Come the spring, the return of the sun, and even before the break-up of the ice, the phytoplankton immediately starts to convert that rich accumulation of nutrients into more microscopic plants. This phytoplankton explosion in turn supports vast hoards of tiny animals, the zooplankton. So the

Walrus, regular visitors to the waters around Coburg Island. (Photo: A. J. Gaston).

web continues: the zooplankton supports arctic cod, which the seals and Brünnich's Guillemots exploit, and the polar bears eat the seals.

When it was calm, before setting off for the colony, we would hang over the side of the Zodiac and gaze into the icy waters. Virtually every few days there would be a new flush of animal life, with one species replacing another through the season. One morning the water was full of sea-ravens, a small (6 mm), highly modified gastropod snail, whose 'foot' has expanded to form two large 'wings', enabling it to fly gracefully through the water. It's shell is tiny, totally transparent and so fragile that even lifting the animal from the water may result in its collapse. Under each black 'wing' there is a fluorescent blue patch which flashes as they flap along. Great shoals of them moved through the water like flocks of slow-motion corvids. We were not the only ones to have noticed them, for huge numbers of Kittiwakes sat on the water's surface furiously dipping and plunging. We did not know what they were eating until we caught some Kittiwakes to weigh them, and they vomited up the remains of their black, sticky meals. The smell was both pungent and tantalizing, and I realized that with a touch of added garlic it would have resembled a French delicacy .

Another, more beautiful pteropod *Clione* had lost its shell altogether. It was about 4 cm long and resembled a golden dagger, with hilt-like wings. Comb-jellies, or sea gooseberries as they are sometimes called, were also abundant, and their iridescent combs held us spell-bound as they chugged slowly through the water. We saw animals that I had told my students about, but had never seen alive. These included medusae, crab larvae and enormous, brilliant red relatives of the comb-jellies. The latter were bell-shaped and 14 cm long, with rows of tiny cilia which propelled them aimlessly through the sea. I caught one to have a closer look, but once removed from its watery crutch it was reduced to an amorphous blob of jelly. On being returned to the sea, it resumed its exquisite form. Our departmental museum contained several jars of these organisms, but despite the best efforts of those who prepare these things for display, comb-jellies and their relatives preserved in spirit medium are about as exciting as blobs of nasal mucous.

* * *

Towards the middle of our field season we were to be visited by a CBC film crew, who were making a documentary to illustrate the conflict between industrial activities of companies like PetroCanada and the conservation of the Arctic wilderness. We fully supported the idea of the film, and I was looking forward to their arrival, not least because they had agreed to bring my wife, Miriam, with them. But I also had some reservations. As an all-male team we had worked well together so far, but I was concerned that Miriam's presence might disturb our equanimity. I was especially sensitive to the potential (and realized) benefits that being in charge could offer, and knew that the unfair distribution of 'resources' could spell trouble. As it was, I need not have worried for there was no overt problem on this front: there was no (apparent) disruption and Miriam fitted in well, playing a central role in the work and camp life.

As anyone who has been part of a small, isolated community will know, it is easy to become terribly territorial and resent the intrusion of new personalities. Because of this, the idea of a seven extra people sharing our Parcol even for just a few days was a cause for some concern. Film crews are notorious for their inflated egos and insensitive ways, behavioural traits which are no doubt reinforced by exactly the sort of instructions we had been given about how we should treat them. We were effectively told to pander to their every wish in order that they could complete their filming with the minimum of hassle. This would have been fine if we had not been struggling against the elements to secure our own rewards from Coburg. I had visions of losing more days of data whilst ferrying film people to and from the colony. Hence my apprehension when I heard over the radio that they had just left Resolute and were on their way.

Because of the changeable weather conditions they were 'on their way' for several days, and ended up being dropped at Truelove Inlet on the north coast of Devon Island. Although Miriam and I were keen to see each other, her enforced stay at Truelove was a memorable one, allowing her to get within a few metres of Musk ox. Indeed, this region of North Devon is one of the most lush and productive areas of the Arctic. It was partly because of this biological richness that Frederick Cook and his two Inuit assistants were able to over-winter here in 1908–1909, living off little other than their wits.

Frederick Cook's enforced stay on the shores of north Devon was part of one of the greatest of all Arctic sagas. On 21 April 1908, Cook and his two companions, Etukishook and Ahwelah, had reached the north pole. Later that same year Robert Peary, with a combination of a much larger entourage and military strategy, also claimed to have reached the pole. Curiously though, Peary did not announce his success until he had heard of Cook's claim. Controversy raged, and still does, but by using every dirty trick in the book Peary managed over the succeeding years to discredit and destroy Cook totally.

No-one knows whether either of them reached the pole. Notwithstanding, Cook's journey from Anoatok in west Greenland, across Ellesmere Island, north along Axel Heiberg to the 'big nail', and back down into Jones Sound and Devon Island and eventually back to Greenland, was a far greater achievement than anything Peary ever did. On his return from the pole Cook and his companions arrived at Cape Sparbo in early September, the beginning of the Arctic winter, with just four rounds of ammunition, half a sledge, a canvas boat, a torn tent, some matches and a few pots, pans, and knives. Their overwinter home was a cave in an abandoned Inuit settlement and they survived by hunting musk ox with lassos fashioned from sledge thongs, and with spears made from their hickory sledge. For 5 months they lived like cave men, fighting off polar bears which tried to steal their precious food and waiting for the sun (Cook 1911; Eames 1973).

The 5 days that the film crew and their various hangers-on were with us were fine. Conditions were crowded in camp, but most of the time they were perfectly reasonable human beings. We had only two minor criticisms. The

Twin otter on the beach beside the camp at Coburg Island: departure of the film crew.

first was what I think might be a genetic defect: none of the film crew knew how to wash up. They were not even shamed into it by Don Gamble, one of the most eloquent and committed of Arctic conservationists I have come across, who accompanied them. The second criticism involved a rather more irksome incident. We were coming back from our study plots late one afternoon. There was a strong breeze from the north which meant that the Zodiac had to punch its way through the waves, and we all got cold and wet. As we rounded the north end of the colony there was an ethereal light among the ice, and the film crew were filming in our direction. With undisguised irritation, they signalled for us to get out of the way, which meant returning the way we had come. We returned to the study area and waited. When we finally got back to camp several hours later, none of the film crew referred to the incident nor asked us how we had killed the time.

On the morning of the sixth day a Twin Otter whisked them all away and with undisguised relief we waved them farewell. We had cleared a landing strip on the beach beside the Parcol, and fortunately it proved to be just long enough and to have a sufficiently suitable surface for the plane to land on. This meant that the film crew were able to leave with dignity, the alternative being for us to ferry them and their several tons of equipment across the bay to the far side of the Marina Peninsula, where we had landed at the beginning of the season. The end result of all this was 'Arctic Oil'—a hard-hitting television documentary, which may have played some role in preventing the exploitation of mineral resources in Lancaster Sound region, at least for the time being.

* * *

Meanwhile, our colleagues at Cape Hay had not had an uneventful season. They remained cheerful despite the persistent bad weather and other demoralizing forces. We spoke to them on the radio each evening when all the other 'traffic' had been completed. It was always good to talk to them, but having to follow the roger-over-foxtrot jargon and the strict code of radio conduct was often frustrating. On one occasion soon after we arrived we naively enthused to them about something we had seen, throwing in the odd profanity for emphasis, but we were swiftly reprimanded for stepping beyond the bounds of acceptable radio-speak. Swearing or blasphemy on the radio was a definite no-no. These rules exist because literally anyone can tune in to these frequencies and eavesdrop on a radio conversation. We enjoyed listening to the conversations between other camps (particularly those involving the female voice at Borden—whoever she was), and I presume our radio conversations with Cape Hay sometimes bemused and entertained others.

One particular conversation I had with Erick remains very vivid. When I greeted him and asked how things were going he sounded positively dejected. He was reluctant to talk and then, quite out of character, asked if his team could have the following day off. I agreed, knowing there must be a good reason for such a request. I then gently tried to persuade him to tell me why. By this stage, Keith, Don and Bill had moved next to the radio, so as not to miss anything through the often poor reception. Eventually Erick disclosed that they had had a particularly bad day with a polar bear. Larry's incident the previous year should have forewarned us, but it was now apparent to Erick, Bruce and Gordon that the plateau-like landscape of Cape Hay was a regular route for bears moving inland for the summer.

This is how he subsequently (and with the benefit of reduced adrenaline levels) described events of that morning to me:

> I had a blind (hide) about a mile to the east of camp, at the end of a pinnacle right along the edge of the cliffs. The blind was perched about 2 feet from the edge of the vertical 1500 foot drop. I had talked to a friend who was studying Ivory Gulls, and he told of a polar bear that had approached a blind without the occupant realizing it. Apparently the bear knew that there was something chewy and delicious inside, and pushed its body up against the front. All of a sudden, everything went dark inside the blind, and then it started rocking violently to and fro. The last thing I wanted was a bear rocking my blind back and forth on the edge of a 1500 foot cliff, so I drilled some peepholes in the sides of the blind. I would alternate glances back and forth between the murres (guillemots) and the tundra behind me. Sure enough, about a week after I drilled the holes there was a male bear about 30 feet from the blind and advancing quickly. You have never seen anyone get out of a blind so quickly! The bear put its head down, ears back, and swayed his head back and forth (which is what they do just before they charge). I gave the bear one warning shot, and fortunately that startled him enough so that he backed up and left!

A day off after such an incident seemed little enough time to recover one's composure. However, their bear troubles weren't over.

Like us, Erick's team had rented a husky from Resolute to serve as an early warning system, and in fact their dog was the mother of our Bruce. (Rather confusingly, we had named our dog after Bruce Llyon in Erick's camp). If a bear had ever come anywhere near our camp on Coburg, Bruce would have been the first one under the Parcol. Luckily for the Cape Hay team, their dog—Beauty, as she was called—was made of sterner stuff.

On one occasion Erick was returning to camp in blizzard conditions, when Beauty suddenly 'stiffened up' and ran off into the fog. A few metres further on Erick came across Beauty's tracks superimposed on those of a large bear. Beauty was chasing the bear, but unfortunately she was chasing it in the wrong direction. The bear ran towards a part of the colony where Gordon and Bruce, in blissful ignorance, were weighing Guillemot chicks. Erick ran off into the fog following both sets of tracks. As he came over a hill top he was able to look down the scree slope below the fog and see the bear rapidly approaching Gordon. Bruce was dangling off the end of a rope weighing chicks and Gordon was looking over the cliff with his back towards the approaching bear. Erick fired a shot in the air to warn Gordon, but because the bear was hidden behind a ridge of rocks he was unable to see it. However, after watching Erick's wild gesticulations for a few moments he did realize that something was seriously amiss. Bruce scrambled back up the cliff just as the bear appeared—fortunately they had a gun and by firing more warning shots they were able scare the bear back up the slope. Shredded neurones.

As though that was not enough, towards the end of the season they were woken early one morning by the shrill cries of Beauty. Looking out they were horrified to see, just a few meters away, a bear mauling the dog. On opening the Parcol door, the bear lost interest in the dog and rushed at Bruce, who in self defence had no option but to shoot the bear. The inch-long lead slug from the shot-gun cartridge did its job, stopping the bear in its tracks. Whenever an incident like this occurs the local community is entitled to the carcass. An old Inuit man was duly shipped out; he spent a full half hour sharpening his knife and then within 1 hour had reduced the bear to neat piles of skin, bones and meat.

* * *

Polar bear early warning system: Beauty at Cape Hay. (Photo: E. Greene).

We noticed that the sun was now slipping behind the mountains behind our camp each evening. As it did so, the night time temperature plummeted and there was often ice on our stream in the morning. The brevity of the Arctic summer was remarkable: it hardly seemed any time since the first eggs were being laid, now the first chicks were about to fledge and summer was rapidly degenerating into winter.

The process of fledging at Coburg was fraught with difficulty for the young Brünnich's Guillemots. The irregular topography of the cliffs meant that many of them did not have an unobstructed passage down to the sea, and to make matters worse there were often huge pans of ice on water below the cliffs. From the colony we would watch young Brünnich's Guillemots jump from their ledge where they had been secure for the past 3 weeks, only to clip a rocky outcrop and die during their descent. Other chicks which avoided this particular hazard would attempt to steer so that they landed on the sea rather than ice, but many failed and died on hitting the ice.

On one particular night in late August when approximately half the chicks in the colony fledged, we watched from the water and in just a few hours we filled the Zodiac with the corpses of those chicks that died. We collected a total of about 500 dead chicks that evening. Although this seemed extreme, we reckoned that it represented only 2% percent of the chicks that left the colony that day.

* * *

Our time at Coburg rapidly came to an end, and because of our teaching duties back in England, Miriam and I had to leave a couple of weeks ahead of the others. It was a sad moment when we said 'goodbye' and our Twin Otter, piloted by 'drunken Duncan' carried us away from Coburg. Within seconds of being airborne we were able to see the glistening tower of Princess Charlotte Monument to the east, and the beginning of the Brünnich's Guillemot colony below us. As we flew along the length of the colony I was able to clearly see the tiny speck of my plywood hide on the cliffs. I was also able to see the route by

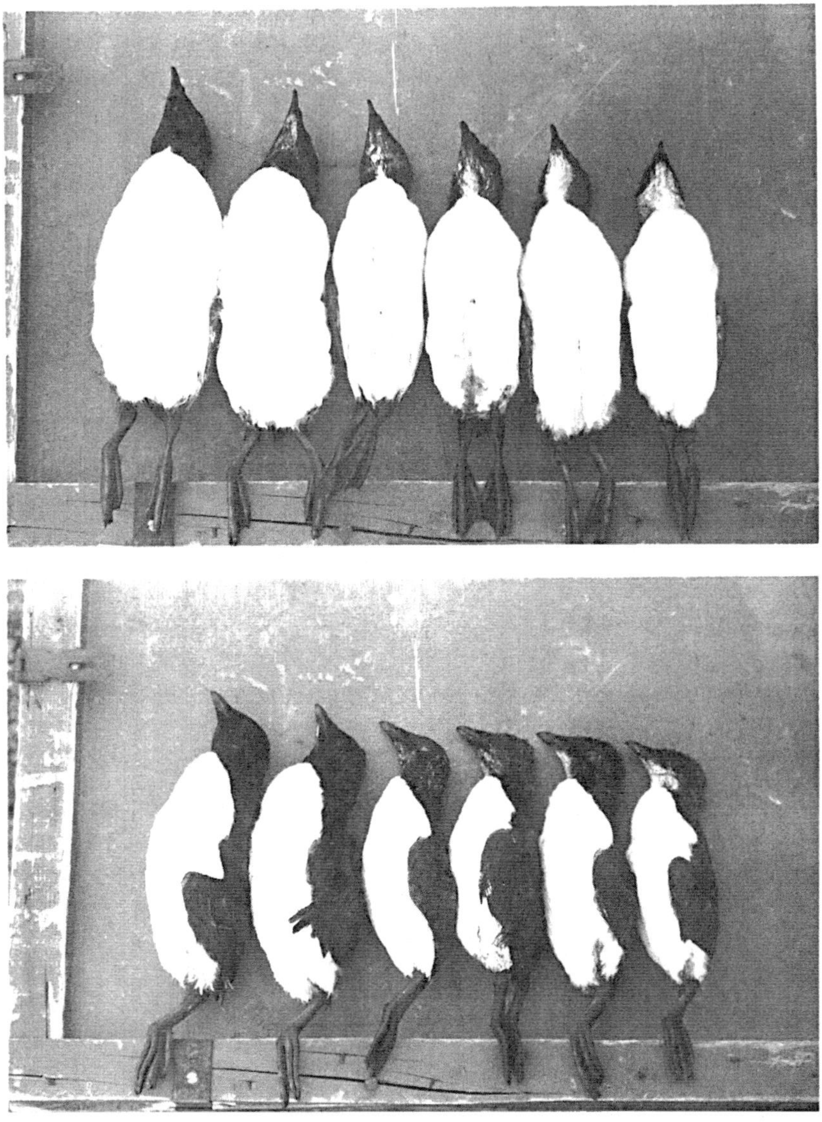

Some of the Brünnich's Guillemot chicks that died during fledging at Coburg Island to show the variation in head plumage. These chicks were arranged in order of size by chance, the plumage variation is independent of size. (Photo: K. Clarkson).

which Don and I had climbed out of the colony and in the fleeting seconds as we flew along the cliffs it was all too clear that there was virtually nowhere else in the entire length of the colony where that escape would have been possible.

As we flew away from Cambridge Point and into Jones Sound, Duncan and his co-pilot started to quiz us about Coburg, and what it had been like. They then asked whether we'd seen musk ox or polar bears. Before I could answer we were in a steep dive over the north Devon coast, and with my stomach still several hundred metres above the rest of me, we circled, on our wing tip over a herd of musk ox which had formed themselves into a classical, defensive circle. I was pleased to be able to say that I had seen musk oxen, but the angle at which we viewed them prevented my brain from functioning sufficiently well to form a proper image. Later in the trip we were more successful with several different polar bears and flying only a few metres above the ice, we were able to view them without resorting to stomach-wrenching aerobatics.

We then rose to several thousand feet, and flying along the frozen length of Jones Sound, we headed for Resolute and home. As we droned forward, Duncan drank a can of Coke. After he had finished, he simply opened the window and chucked out the can. In that one thoughtless action he exemplified all the problems faced by remote places.

CHAPTER 4

The Lives of Great Auks

Some ornithologists have indulged in the hope that in some hitherto unexplored part of the Northern Seas we would yet find the living Garefowl.

Grieve (1885)

On the west coast of Newfoundland's Great Northern Peninsula at Port au Choix, lies a prehistoric burial site 3000 to 4000 years old which belonged to the Maritime Archaic people, the first known occupants of Newfoundland. Excavation revealed many small, finely polished, white pebbles, part of a rich legacy of relics demonstrating this people's adaptation to the coastal environment. It is clear from their artefacts that the Maritime Archaic people subsisted on marine mammals, seabirds and fish, and that birds held a special place in their ceremonial lives. By 500 B.C. the Maritime Archaic people were extinct.

Funk Island, a small, low-lying granite rock, lies 60 km off Cape Freels on the north-eastern coast of Newfoundland (Fig. 6). With myriads of seabirds in the air above him, a biologist kneels on a small patch of turf, scraping gently at the peaty soil with a knife. A small, almost spherical pebble, about 1 cm in diameter, rolls out of the earth where he is digging. He picks up the pebble, spits on it and rolls it in his hand to clean it. He carefully places the whitish pebble with several others which he has just found: they are the gizzard stones from Great Auks.

* * *

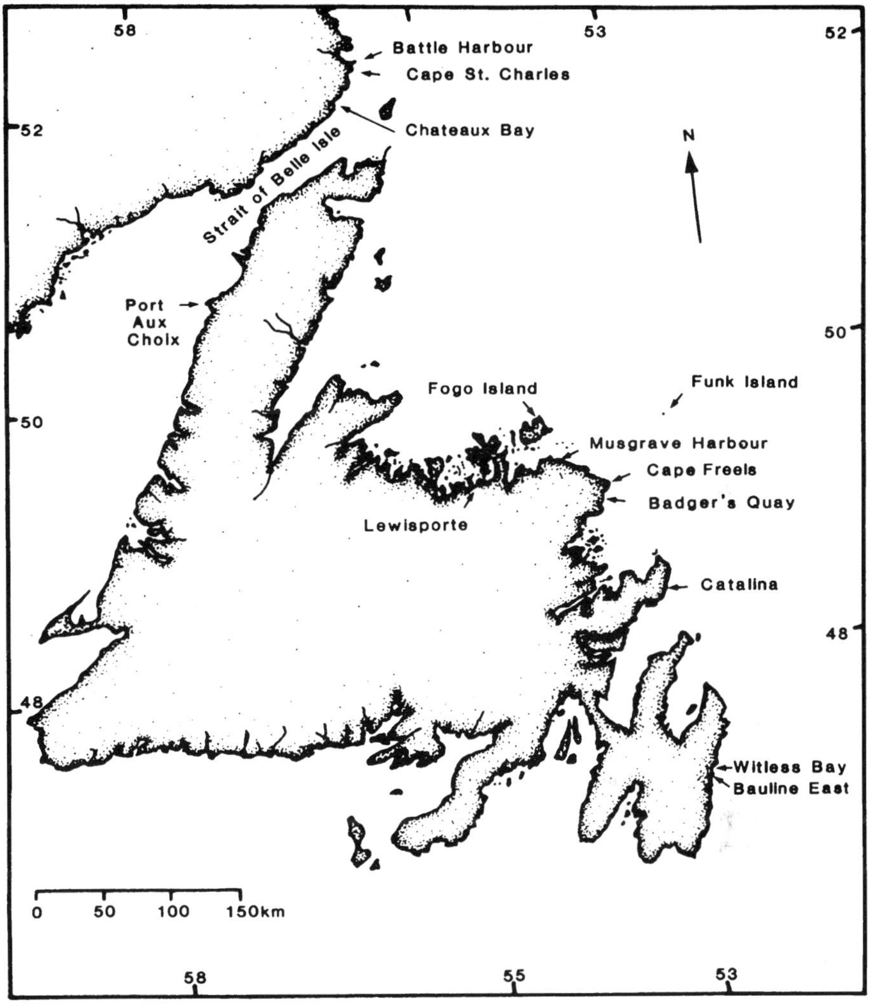

Fig. 6 *Map of Newfoundland showing locations mentioned in the text.*

Three hundred kilometres to the south, in Witless Bay are three seabird islands: Great, Green and Gull Island. The largest is Great Island, the breeding home of about 150,000 pairs of Puffins, as well as Common Guillemots, a few Razorbills and many thousands of Herring Gulls. This was an unusual seabird island for me, differing in several respects from most others I had visited. First, it had a good covering of trees, mainly black spruce, which in the centre of the island formed a dense, if diminutive forest. Even more incongruous, the numerous Herring Gulls rested in the trees, like so many crows, albeit wobbly ones since their webbed feet did not permit them a firm grip. Adjacent to the forest there was meadow-like habitat: lush, green, knee-high grass patrolled by an abundance of two species of swallow-tail butterfly (the Eastern Black, and the Eastern Tiger). The forest was also the home of thousands of Leach's Petrels, among the smallest and gentlest of all seabirds. There is a curious history to this bird's name. The mounted skin from which the species was identified was part of William Bullock's museum, based initially in Sheffield at the end of the 18th century. Bullock himself had, in 1812, taken part in a six-oared boat chase after one of the last Great Auks off Papa Westray, Orkney. The bird (a male) escaped, that time, but it was killed the next year and took its place in Bullock's museum. Six years later however, in 1819 Bullock decided to sell his collection, and this is when Leach bought the petrel (together with the Great Auk and its egg) for the British Museum. Later the Dutch ornithologist, Coenraad Temminck (of Temminck's Stint fame), saw the specimen in the British Museum and realizing it was a new species, named it in Leach's honour (Mearns and Mearns 1988). Lucky Leach—by rights perhaps we should be referring to Bullock's petrel!

Leach's Petrels breed in underground burrows and come ashore only at night since their small size (they weigh just 50 g) renders them vulnerable to predators, such as gulls. In the forest they excavate holes in the soft humus, often between the roots of trees and it was sometimes possible to hear their melodious crooning from within their burrows. Like all members of the petrel family, Leach's has a distinctive and persistent body odour, thought to emanate from their plumage and ultimately from their large preen gland. Those portions of the brain concerned with detecting smell, the olfactory lobes, are also relatively large in petrels, and it is thought that they find food (as well as their burrow) by smell, and they can probably also recognize their partner's scent.

As I walked from one side of the island to the other I always relished going through the forest. There were no gulls, and the penetrating musty smell and extraordinary calls of the petrels, combined with the redolence of the trees evinced a Breughelesque world. The petrels also bred in the vicinity of the cabin and at night they fluttered through the beam of light cast from the window like an endless stream of moths. These beautiful birds were the tranquil sanity on an otherwise aggressive island. If they had not been nocturnal and burrow-dwelling and hence unwatchable, I might have been tempted to study them.

Great Island's terrain was steep and irregular and the highest points provided spectacular views, both to the other islands and to the mainland. Like all islands, this one's inaccessibility made it seem remote, yet the sight of cars cruising up and down the highway on the adjacent mainland was a constant, and for me at least, unpleasant reminder of the proximity of what we call civilization.

Great Island is where in the 1960s David Nettleship (1972) had carried out his PhD studies on the effect of the island's large population of Herring Gulls on the Puffins. This is a story of good guys and bad guys, with the Puffins being the good guys. The number of Herring Gulls on both sides of the North

Atlantic has increased enormously during the present century, and part of this increase is thought to have occurred as a result of human activities. By providing an abundant winter food source in the form of our refuse tips, we have made it easier for Herring Gulls to survive the winter, and as a result their numbers started to increase. Large numbers of gulls at their breeding colonies make life difficult and unpleasant for other seabird species, researchers, and for themselves.

David's study revealed that gulls made life difficult for at least some Great Island Puffins, mainly through stealing fish that the Puffins brought for their chicks. The robbing of one species' food by another is referred to as 'klepto-parasitism', and in this case resulted in Puffin chicks running short of food. Puffins are rather poor flyers and certain 'specialist' gulls learnt to exploit their locomotory shortcomings. The gulls' high speed aerial attacks on birds returning to the colony with fish look pretty brutal: if the Puffin gets as far as alighting on the cliff top the gull will half run, half fly at it and grab it in its bill. The Puffin is then shaken until it releases its load of fish, which it usually does very readily. Once the fish are dropped the gull has no further interest in the Puffin, and releases it in order to quickly snap up the discarded fish. Some gulls start to pursue Puffins in the air as they approach the colony. When this happens the chase may last many minutes, with Puffins veering off back out to sea, with one or more gulls in hot pursuit. As the gulls close in the Puffin will drop onto the sea and dive. However, there is no escape, because as the Puffin surfaces the ever-agile gulls fly over to it, causing the Puffin to dive again. The gulls are persistent and there is a limit to how many times the Puffin can dive without catching its breath. The end is inevitable: the Puffin abandons its catch to the gulls. This must be something of a disappointment for the Puffins, since, judging from the relatively low frequency with which they feed their chicks, food must be hard to come by. However, the Great Island Puffins should be thankful for small mercies, at least they are not killed by gulls, as they are at some other colonies. At St Kilda for example, some Great Black-backed Gulls specialize in rearing their own chicks on a diet of adult Puffins.

Not all Puffins on Great Island are equally vulnerable to kleptoparasitism by gulls. Those with burrows on sloping terrain near the cliff-edge avoid such harassment since they are able to fly more or less directly in and out of their burrows faster than the gulls can respond. On the other hand, the Puffins with burrows on level ground away from the cliff-edge are especially vulnerable to gull attacks. This is because they usually have to alight on the cliff-top and then walk (or run if there is a gull about) to their burrow. Overall, Puffins on level ground lose about 13% of their chick meals to gulls compared with just 4% for the Puffins on sloping ground. The result of this is that Puffin chicks on level ground grow more slowly and survive less well than those on sloping ground. Puffin chicks which are short of food come to the entrance of their otherwise safe burrow, anxiously awaiting the return of a parent with food. However coming to the burrow entrance simply makes them more likely to be killed and eaten by a gull. The proportion of Puffins that successfully raised chicks to the fledging stage was twice as high for those breeding on sloping

ground (0.5 chicks per pair on average), compared with those on level ground (0.25 chicks per pair).

The sight of a gull viciously shaking an innocent-looking Puffin isn't endearing, but gulls treat their own species with similar contempt. Herring Gulls are not fussy whether they are feeding their offspring on Puffin chick or Herring Gull chick, just as long as it is someone else's. In a gull colony it's dog-eat-dog and any unprotected eggs or chicks are rapidly eaten by other colony members. This sounds repugnant, because it is contrary to what we might expect. It is just about okay to eat the offspring of another species, but should members of the same species not be looking after each other? That is what we have been taught to believe from our earliest schooldays, but, it misses the point about the way natural selection works. There is nothing sacrosanct about one's own species: if by cannibalizing its neighbour's chicks a Herring Gull can rear more of its own offspring than a non-cannibal, then the genes for cannibalism will tend to spread through the Herring Gull population. The important point is that Herring Gulls do not eat their own offspring.

So Herring Gulls make life difficult for Puffins and other Herring Gulls, but what about the innocent researcher? On Skomer Island I had to walk through a mixed colony of Herring and Lesser Black-backed Gulls every morning to get to my study area. The colony was so extensive that it was virtually impossible to avoid it. Not unexpectedly, the gulls resented my presence and treated me as they would a predator, by diving at me and generally making a lot of fuss. Their discordant cries and rushing wings never bothered me—until I was hit. When a gull dives towards you it emits an evil, sneering screech which reaches its crescendo just at the point when the gull is closest to you. I had never thought too much about this call, until my first physical contact with a gull. On this occasion the gull caught me completely by surprise, hitting me with both its feet on the back of my head. As it did so it managed to defecate and vomit simultaneously, with the result that my head was covered in a large quantity of brown sludge the consistency and colour of mud, but with a smell that was far worse. My reaction was a mixture of anger, revulsion and a horrible sinking feeling in my stomach which was the result of a kilo of gull unexpectedly hitting my head.

The gull's anti-predator behaviour has evolved through natural selection to produce the most effective deterrent at minimum risk to itself. The behaviour comprises a rapid aerial attack, an unnerving screech followed by a discharge of semi-solids. By swooping down at high speed the gull combines intimidation with an ability to avoid capture by the predator. The high speed deposition of faeces is undoubtedly a modified form of what was originally a fear response and is effective because all animals dislike any contact with fresh faecal material. This is not simply an aesthetic response. Our avoidance of faeces no doubt has a solid biological basis which is the avoidance of pathogens, such as bacteria, parasites and so on.

In one bird species defecation as a hostile act can have more direct consequences if the attack is concerted enough: mass defecation by colonial Fieldfares can soil a raptor so badly as to ground it and sometimes even lead to its death. Fortunately gulls have not specialized to this extent.

But what about the gull's awful call? This is the icing on the cake, so to speak. I suspect that the call was added because it so increased the effectiveness of the gull's attack. Much as Pavlov's dogs associated ringing bells with food and resulted in their salivating, I (and no doubt other would-be predators) now associate the gull's screech with capital abuse, and the sound of the screech alone makes me take avoidance action whenever possible.

In fact, unless one went looking for trouble with gulls on Skomer, they were remarkably tolerant, and relatively few people were ever hit in the way I have described. Great Island, in Witless Bay, was however, another story. I have never encountered such aggressive Herring Gulls anywhere. Not only did they follow-through a high proportion of dives with a direct hit and defecation, they also used their beak and feet to inflict deep scalp wounds. This was not a case of my being a wimp; a single trek across the island without a stick to deflect attacks could be a nightmare. Not only were there gulls virtually everywhere, but the ground was riddled with an incredible density of Puffin burrows so that walking was extremely difficult anyway. One only had to glance upwards in anticipation of a gull attack to stumble in one or more burrows. To add insult to injury it rained heavily virtually every day making the ground sodden and slippery, and walking, even without gulls, was tricky. My memories of Great Island are very mixed.

* * *

In Witless Bay seabirds and men compete cheek by jowl for the same rich harvest in the sea, but their coexistence is an uneasy one. Looking into the shallow water below the wharf at Bauline East I could see myriads of small, black sea urchins, their population fuelled by grazing on countless cod guts. Also on the wharf were the corpses of a dozen seabirds: several Puffins, Common Guillemots, a Razorbill and a Black Guillemot. They had been drowned in cod gill-nets set near the seabird colonies. The number of dead birds in the pile could easily have been dismissed as insignificant, but it was part of a much larger problem, and adding up the numbers drowned for each locality, for each day, for each month and for each year, demonstrated that many thousands of birds, mainly Common Guillemots and Puffins, die annually (Piatt *et al.* 1984). A single organism, the capelin, is at the root of this wasteful mortality.

The capelin, a diminutive and distant relative of the salmon, is a key species in the marine food web of Newfoundland and Labrador waters (see Chapter 7), and in other low Arctic regions, such as Iceland and southern Greenland. Predatory fish, marine birds and mammals, such as whales and seals, depend upon it almost entirely at certain times of the year. Although Newfoundland people have exploited capelin on inshore spawning beaches on

Capelin drying in the sun at Badger's Quay, Newfoundland.

a small scale for many years, it was not until 1972 that a large commercial offshore capelin fishery started, to cater for the Japanese who enjoy capelin roe. As such the fishery was extremely wasteful since only ripe females were used; males and unripe or immature females were simply discarded (dead). The initiation and development of this fishery followed a pattern that is becoming increasingly familiar.

The basic pattern is this: overfishing of large, predatory fish like the cod, results in poorer and poorer catches consisting of smaller and smaller fish. A reduction in the number predators (such as the legendary Newfoundland cod fishery achieved) results in an increase in the number of their prey, the capelin in this case. This in turn means that seabirds have a relatively easy time, and their numbers also start to increase. It generally takes a while for the fishermen to realize that the tiny prey species are commercially worthwhile, but once they do, the effect of their fishing can be devastating, and not just for their prey. The landings of capelin in Newfoundland rose from 20–25,000 tons to a peak of 370,000 in the mid 1970s and then declined markedly. In an effort to safeguard the fish stocks the offshore capelin spawning area on the southeast shoal was closed to fishing in 1979. But the same year, in a curious clash of loyalties, the inshore fishery was actually expanded in an attempt to safeguard the fishing industry. The years 1979 and 1980 were marked by a widespread spawning failure of capelin (Carscadden 1984).

The establishment of the capelin fishery in Newfoundland waters coincided with several major ecological phenomena. First, within a few years of the start of capelin exploitation very large numbers of baleen whales, mainly humpbacks, started to occur in inshore waters. The numbers of whales seen

Humpback whale below the surface. (Photo: Worldwide Fund for Nature).

was unprecedented, and for anyone interested in natural history their presence could provide awe-inspiring spectacles. One afternoon on Great Island I was sitting on the cliff-top watching flocks of Common Guillemots returning to the island. All the Common Guillemots from the three Witless Bay islands seemed to have been feeding in the same general area and the steady stream of birds flying in towards the islands was remarkable. As I stared at the black lines of birds above the distant horizon I became aware of another kind of movement in the sea: huge numbers of humpback whales were spouting and lunging upwards out of the sea, presumably feeding on the same capelin shoals that the Common Guillemots were enjoying. I tried to count the spouts, but gave up somewhere around a hundred.

A few days later as Anne Greene and I were counting the birds on Great Island's cliffs, a solitary humpback broke the surface just a few hundred metres offshore. It then proceeded to breach no fewer than seven times in succession. We were both excited, and I wanted to watch the whale through my binoculars, but each time it leaped from the water Anne grabbed my shoulder, excitedly shouting 'There it is again!'. I never got to see it through binoculars. Although we enjoyed seeing, and later being among, the whales on the sea, their effect for local fishermen was often disastrous as they became entangled with commercial fishing gear. Whales died and fishermen lost large amounts of expensive equipment. In 1979 for example, 13 humpback whales died and the fishing industry lost $2.5 million through whale damage. Subsequently Whitehead and Carscadden (1985) were able to show that the whales moved into inshore waters because their capelin prey was scarce offshore. The implication that man's activities had reduced capelin numbers seemed obvious.

Humpback whale lunging out of the water showing the folds of its mouth. (Photo: K. Balcomb).

The second phenomenon occurred in 1981. In that year almost the entire production of 100,000 Puffin chicks on Great Island starved to death. The few that survived to fledge were seriously underweight. The reason for this reproductive failure was simply that there were no capelin. In fact there was little food of any sort because there was virtually no alternative prey for the Puffins to switch to. Those adult birds that did find food for their chicks brought back small fish of the cod family, renowned for their low energy content. The breeding success of Puffins in that year was virtually zero whereas in previous years at least half of all pairs successfully reared a chick. The sight of thousands of dead or dying Puffin chicks can trigger strong emotions, and one's first reaction might be to blame the greedy capelin fishery (Homer 1982). However, the fact that in the years following 1981 Puffins were able to find capelin and rear their young successfully, despite the continued exploitation of the inshore capelin stocks, means that the relationship between man's fishing effort, capelin abundance and the Puffins' breeding success is not a simple one (Nettleship 1991).

The events in Newfoundland with the capelin fishery and seabird breeding failure have also occurred elsewhere. In the Barents Sea off northern Norway, for example, the lack of capelin has resulted in massive declines in Common Guillemot numbers (Anker-Nilssen and Barrett 1991). In the North Sea there exists an almost unregulated sandeel fishery, with an annual catch of over one million tonnes. The Danes take most of this, which is not even used for human consumption, but instead is used to fuel Danish power stations and pigs. A sandeel fishery started in Shetland waters in 1974, and landings increased to a peak in 1982. Shetland's sandeels are processed into fish meal and are used to feed farmed salmon and to make fertilizer. From about 1984 until 1990 Shetland's terns and Kittiwakes showed consistent breeding failure, with their chicks starving to death year after year. Public pressure resulted in the fishery being closed prior to the 1991 breeding season, and in that year the breeding success of the birds showed a remarkable recovery (e.g. Monaghan 1992). The obvious conclusion is that the closure of the fishery resulted in the change of fortunes for the seabirds but it could have been due to some other factor. Unfortunately, because fish populations are difficult to census, fisheries biologists have little real idea of fish abundance, which means that the concept of managing fisheries is either optimistic, naive or erroneous (or all three). The evidence for this is overwhelming. Moreover, their attitude offers little cause for optimism: a senior fisheries biologist was heard to say at a meeting that it was 'impossible to damage a marine ecosystem by overfishing, one could only change it'.

* * *

We were in Witless Bay to set up study plots for Common Guillemots on Great Island, and also to attempt to count the number of both guillemot species on Green Island, about 8 km to the north. The carrot that had drawn me to Newfoundland however, was the chance of a trip to Funk Island, a Mecca for seabird biologists. I was also interested in seeing Green Island for myself. A photograph of Green island had featured in Les Tuck's book on guillemots, and he had identified it as one of the few places where Common and Brünnich's Guillemots bred side by side. I had therefore given the island special status in my mind, and in total ignorance of its physical features, even imagined that this might be the place to conduct a comparative study of the two species. The reality was rather different: Green Island is small (510 m × 240 m), and rises abruptly out of the sea to 54 m, with no easy landing place. Nor is there any water, other than rain water, on the island. Certainly there are plenty of seabirds there, but it is not the sort of place to stay unless you have to. In 1973 a graduate field assistant censusing seabirds for the Canadian Wildlife Service had got stuck here because of bad weather and escaped by the skin of his teeth. He had gone for a 2-day visit and ended up staying somewhat longer. A helicopter tried to rescue him but was unable to land because of the danger of collision with the huge numbers of birds. Fortunately it dropped some food, and returned to pick him up after he had spent a total of 11 days on the island! Two days after his rescue the storm was still so bad that all the fishing boats at anchor in the local harbour sank! I was told this story as we attempted to land on Green Island, but it was not until several years later when I met Bob Montgomerie (an established behavioural ecologist), that I realized his previous claim to fame.

Getting to and on to Green Island was an adventure. The seas were full of humpback whales, and we saw them on every trip we made. On one occasion, mid-way between Great and Green Islands, a mother and calf surfaced and blew just a few metres from our Zodiac, leaving us ecstatic. Despite their vast size and proximity, the whales never felt like a threat, instead we almost felt as though they had a benign regard for humans. This was naive considering the many humpback whales that have suffered intentional and unintentional human abuse in Newfoundland waters.

The seas around the islands were always rough, so landing from an inflatable boat was never easy and even after having made it to shore, getting onto the flat top of Green island was another problem. There were a few areas where we could have scrambled up, but this would have entailed the disruption of thousands of pairs of Common Guillemots. Instead we chose to jumar up a vertical cliff-face on a rope which had been placed there previously. The climb would have been fine had it not been for the fact that after each rain storm the route up the cliff-face became a miniature waterfall. However, getting wet was a small price to pay for the spectacle at the top. The number of Common Guillemots was breathtaking, especially when one considers that the older fishermen on the mainland can remember when there were none there at all. In fact they were first recorded breeding on Green Island as recently as 1936. Common Guillemots may well have started to breed on these islands before that since we know that they had been abundant in Newfoundland in earlier times, but relentless exploitation for food, bait and oil by European fishermen from the 1600s onwards reduced populations to a very low level. It was not until the beginning of this century that persecution ceased and numbers started to increase. As is typical of population growth of any life form, the increase in numbers was very slow initially, but then occurred rapidly. By

Common Guillemots at Green Island, Witless Bay on the edge of Southern Cove. This is part of colony of over 70,000 pairs and local fishermen can remember when there were no guillemots breeding here!

1941 there were an estimated 3000 pairs of Common Guillemots on Green Island, increasing to at least 70,000 by 1980 (Nettleship 1980). Only when the population had reached several tens of thousands in the mid to late 1950s did Common Guillemots start to colonize neighbouring Gull and Great Islands.

It was possible to discern the pattern of the Common Guillemot's recent expansion on Green Island. The birds had probably occupied the steepest cliffs initially, but as numbers increased birds were forced, or prepared, to use other habitats. On the south side of the island there is a deep gulch, and it appears that the birds have expanded upwards out of this, and are now overflowing onto the flat ground at the top of the cliffs. Vast numbers of immature non-breeding birds fringe the colonies, their activities gradually destroying the vegetation and reducing the substrate to rubble or bare rock. The distinction between the grass and rock marks the edge of the colony and the sight of large numbers of birds standing on grassy areas at the periphery of established breeding groups is a clear sign of an expanding population. During my PhD studies on Skomer I had never seen large numbers of non-breeding birds behaving in this way, which was hardly surprising since the population there was much reduced and was not expanding. More recently, however, as Common Guillemot numbers on Skomer have increased, exactly the same phenomenon has occurred. Guillemots of either species will only lay their egg on sites free of vegetation, but whether the trampling and destruction which goes on in the year or two prior to breeding is a deliberate strategy to achieve this is not clear.

The spectacle of Green Island's Common Guillemots was matched only by that at Funk Island.

* * *

We arrived at Musgrave Harbour, 50 km northwest of Cape Freels, just as a clear, calm dawn was breaking. Anne Linton, David Nettleship and I, in the company of two of Jacques Cousteau's team, were heading for Funk Island, one of the world's largest seabird colonies. Our chartered long-liner, the *Dougie-D*, and her cheerful crew were ready and waiting, and without any of the usual delays we set off. The trip was pleasantly dull: the seas were calm and apart from a few groups of Common Guillemots and some Great Shearwaters there were few distractions. The only blot on this otherwise peaceful seascape was the female member of Cousteau's reconnaissance team: one of those frighteningly aggressive, no-nonsense sort of North American ladies that one occasionally encounters. Fortunately I did not have to interact with her directly, but her arrogant, assertive manner with the others on the boat made me edgy.

After 3 hours the low horizon of 'the Funks' became visible. The name, old English in origin, means foul smelling or place of panic. Foul smelling it certainly is for the slippery guano of a million seabirds has more than a hint of ammonia about it. The panic aspect probably reflects how seafarers felt about the dangerous outlying rocks, but the name also anticipated the fate of the Great Auks breeding there. As the island first came into view I imagined that this was what it must have looked like over a 150 years ago for the Great Auks, or Penguins as they were originally known, returning there at the start of the breeding season.

My impression of Cousteau's lady did not improve during the course of the day. She wandered round the island alone, and it was clear from our moments of conversation when we met that she hated this alien environment. She was ill at ease, and a liability with the birds. The Common Guillemots at Funk Island breed on flat ground in groups of hundreds or thousands and one can walk directly up to them. Superficially they appear to be unperturbed by the presence of people, but in reality they are in a state of desperate suspense. All one can do under these circumstances is to move slowly and carefully and above all give them as wide a berth as possible. She did not do any of these things, and at one point I looked up to see her stumbling directly towards an aggregation of several thousand Common Guillemots. I watched, my heart in my mouth, hoping that she would change her course. In utter disbelief I saw her blunder right into the birds, causing them to flee in panic scattering their eggs and newly hatched chicks as they did so. Even then she did not stop. For her the birds were simply an irritating, inert part of the landscape. Depressed and angry I made my way towards her but I could tell from her face that remonstration was futile: she had had enough of Funk Island.

During the day a strong wind had got up from the south. Most of the island

Part of the central Common Guillemot colony on Funk Island.

was relatively sheltered, but it was apparent from the surf off the south coast that our journey back was going to be less comfortable than our outward one. I took precautions, and slipped two sea-sickness tablets into my mouth half-an-hour before we set off.

Our counts complete, we boarded the *Dougie-D*, which was moored at The Bench on the sheltered north side. By this time the sky was overcast and the wind was whipping the tops of the waves into horizontal spray. As soon as we left the lee side of the island the boat began to lurch and roll in slow motion epilepsy. Despite my precautionary tablets I knew this trip was going to require a great deal of concentration to avoid the despairing nausea of sea-sickness.

The others all looked pretty cheerful and, hanging on to the rigging, they joked about the boat's unpredictable movements, laughing every time a wave broke over us. Everyone who suffers from sea-sickness has their own survival strategy, at least until things get so bad that they no longer care. My approach consisted of staying outside, however cold it got, breathing deeply and not talking to anyone. In view of the almost total lack of progress we seemed to be making I was relieved to discover how effective my tablets were: without them I would have been rolling around on the deck in my own vomit. Gradually the others stopped laughing, or talked less and I noticed that all of them, except the crew, had turned pale, and even a delicate sallow in some cases. Cousteau's lady was where I could so easily have been: curled up on the deck with her head in her lap.

The crew were cheerfully indifferent, swinging a bucket over the side to swill the remains of so many semi-digested packed lunches from the deck. Remarkably, after some spectacular projectile vomiting, my colleagues started to cheer up and were able to talk to each other and compare complexions. One of the crew emerged from the cabin bearing drinks for everyone. I declined. A while later he reappeared bearing a large Kilner jar of what looked like bright pink rhubarb. As reluctant sailors know, the difference between the real world and the misery of sea sickness can be minuscule: a whiff of cigarette smoke or diesel fumes can push you over the edge. I suspiciously watched the crew member encourage everyone to dip their fingers into the jar and pull out great gobs of what I thought was fruit. As my turn came I saw with horror that it was not rhubarb at all but pickled rabbit, a Newfie speciality.

Later we were plied with various spirits, and again I held onto to my hard won lack of nausea and refused. Cousteau's lady was still on the floor, but the others gladly accepted. Up until this point the skipper and his crew had been models of Newfoundland hospitality, but as we lurched into our eighth and final hour on the *Dougie-D*, I wondered whether there was not an element of cupboard love at work. It was only now that they started to negotiate a price for our day's outing. I stood as a spectator, hardly able to believe that the cost of this jamboree had not been negotiated previously. David, still slightly green, but with plenty of pickled rabbit and rum under his belt suggested terms that, judging from the skipper's expression, were somewhat less than he expected the Canadian Government to be able to afford. Faces turned sour and harsh words were exchanged, but the offered price did not change. The last few kilometres were quiet and tense and it was almost dark when we stepped ashore at Musgrave Harbour rather less cheerful than we had arrived.

The next morning in our hotel at Badger's Quay, David, Anne and the Cousteau couple were all on fine form. Immediately after breakfast David and the male Cousteau set off for Gander to pick up their chartered single prop plane so that they could fly back up to photograph Funk Island from the air. A good night's sleep can work wonders for morale, as can a protracted dose of sea-sickness. The Cousteau lady was reborn, she was a new and better person, cheerful, modest and with a sense of humour. At first I was convinced that the transformation had occurred because of the humbling experience the day before, but there was also what biologists refer to as a confounding variable here that I had not considered. Her change of attitude may have been nothing to do with sea-sickness, it could equally well have occurred because she no longer needed to maintain her professional posture since our leader, David, had disappeared.

* * *

The Great Auk is extinct. The last two individuals were killed by men for their skins on a small rocky island off Iceland, in 1844. Once abundant, the Great Auk was avariciously exploited for its meat, eggs and feathers. There were warnings by early conservationists about the damage being done and there was even a law against the exploitation of Great Auks, with severe punishment for offenders. Notwithstanding, the end result was extirpation. There are numerous reasons for grieving about the loss of the Great Auk, but the one I find so discomforting is that we appear to have learnt so little from it: the pattern of events leading up to the final killing has been repeated so many times since.

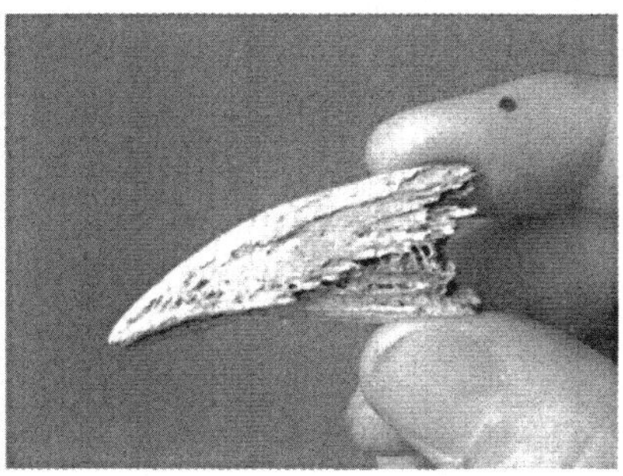

The tip of a Great Auk's beak, unearthed from the soil on Funk Island. (Photo: D. Hollingworth).

The Great Auk was an impressive, flightless bird, standing about 70 cm high and weighing in at about 5 kg, five times as much as a Common Guillemot and almost eight times that of its closest relative, the Razorbill. We can be fairly sure that in the absence of man, Great Auks were long-lived birds.

Considering all bird species, as a general rule the largest have the highest survival rates. This pattern also occurs in the auk family (Fig. 14), and indicates that Great Auks may once have been among the longest-lived birds. Seabird enthusiasts revere the Great Auk as the ultimate member of its family, much as the Raven is among corvid aficionados. It is not just present-day seabird biologists that venerate this species; judging from the abundant remains in their 3000- to 4000-year-old burial chambers, the Maritime Archaic people also worshipped the Great Auk. A human skeleton in one of these ancient graves was found to be covered by no fewer than 200 Great Auk beaks (J. A. Tuck 1975; 1976).

The Great Auk has had a variety of names. It was the original 'penguin', and it has also been known by its Norse name, the Garefowl (or Gairfowl). The origin of this name has excited a lot of speculation and the idea I favour most is Lockwood's (1984): when the Vikings started to settle in the islands north of Scotland in about 800, their encounters with the Great Auk would have left them both amused and perturbed—no other bird they knew was incapable of flight. With a combination of sardonic humour and evasiveness, they named it 'geirfugl' after that master of the air, the Gyrfalcon, literally 'spear-bird ('geir' was the Norse for spear and 'fugl' simply meant bird) indicating the bird's excellence. The name 'penguin' was first used in the late 1500s in Newfoundland and has obscure origins. One suggestion is that it is derived from the Gaelic 'Pen-Gwyn' meaning 'white head', supposedly referring to the two large, oval, white patches on the birds' head. The other, more likely idea is that the name came from 'pin-wing', a reference to the bird's diminutive wings (see Fig. 7). There are several Penguin Islands off the Newfoundland coast, some of which may have been Great Auk colonies. In 1768 Thomas Pennant came up with 'Great Auk', and within a few years the term 'penguin' was reserved for the flightless birds of the southern hemisphere.

After they had become extinct such was the demand for their skins that the taxidermists at the Hancock Museum in Newcastle upon Tyne (and probably elsewhere), spent their lunchtimes making fake Great Auks from the skins of Razorbills. The only tricky bit for the taxidermists was the bill, and they simply carved this out of wood. The wings were no problem, since the Razorbill 's wings were about the same size as those of the Great Auk, despite the Razorbill being a fraction of the Great Auk's size. The Great Auk's wings were tiny relative to its body size (Fig. 7) because they were used only for diving and not for flight. We can assume that the Great Auk's wings had evolved to an optimal size for underwater flight, similar to those of the penguins (see Fig. 7). The volant (flying) auks, of course, had to be able to fly through air as well as through water. As a result they had to have much larger wings. However, during the annual moult most volant auks lose all their flight feathers simultaneously, rather as many ducks do, and become flightless. While this might seem disadvantageous, it actually minimizes the time that the birds are flightless and as Storer (1960) has pointed out, probably makes little difference to their underwater swimming abilities (see Fig. 7). If the Great Auk had followed the same pattern of simultaneous primary moult, this would have reduced its wing area and impaired its diving ability. However, in

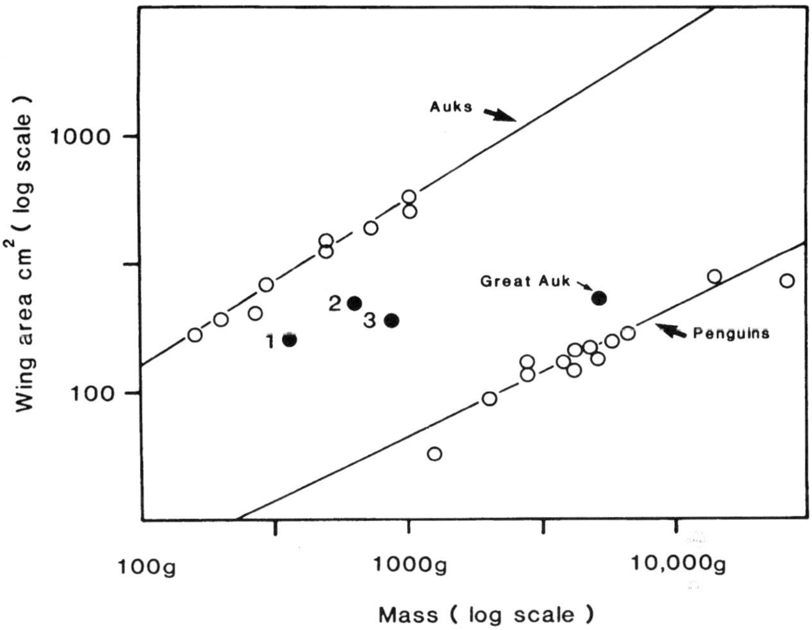

Fig. 7 *Wing size in auks and penguins. The top line shows the relationship between the area of both wings and body weight for nine species of extant and flying auks. The lower line shows the same relationship for 15 species of penguin. The slopes of the two lines are similar indicating that in both groups of birds wing area increases with body size to a similar extent. The fact that the penguin line lies far below the auk line shows that for a given body weight penguins have much smaller wings than auks. Note how close the Great Auk point lies to the penguin line, illustrating the fact that the Great Auk had relatively tiny wings for its body size for an auk, but had a wing area similar to that of penguins. The other points show the wing areas of moulting auks (1. Atlantic Puffin, 2. Razorbill, 3. Common Guillemot), without their primaries. The size of their moulting wings are still relatively large compared with penguins (i.e. these three points fall well above the penguin line). Data for the top line from Livezey (1988), the rest of the data from T. R. Birkhead (unpublished).*

contrast to all the other large auks, the Great Auk actually shed its primaries one at a time (Salomonsen 1945), like most other birds.

What we know about the Great Auk is pieced together from a variety of sources, some based on written accounts, some based on educated speculation. The species was confined to the north Atlantic and we know of at least seven locations where it bred and several other places where it might once have bred. The distribution map of the known breeding colonies (Fig. 8) suggests that this was an auk of low Arctic and Boreal waters with a range overlapping, but not extending as far north as that of the Razorbill. We also know from contemporary records and archaeological evidence that the Great Auk spent its winters far from the breeding colonies, just as extant auks do today. There is evidence of one sort or another that Great Auks travelled as far south as Florida,

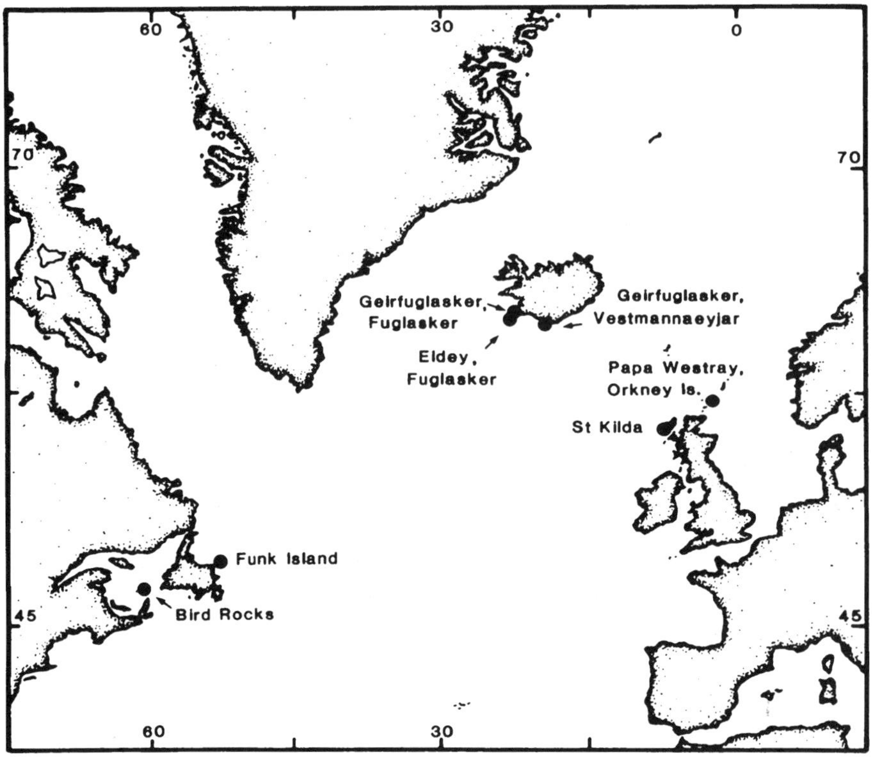

Fig. 8 *Map showing the location of all known Great Auk breeding colonies. From Nettleship and Evans (1985).*

and as far north as Saglek Bay in Labrador and Disko Bay in west Greenland. In the eastern Atlantic Great Auks were seen as far north as central Norway and possibly even Spitzbergen and as far south as Gibraltar and southern Italy.

The observations of Great Auks in Greenland were made by Otto Fabricius who lived at Frederikshab in southwest Greenland between 1768 and 1774. He recorded the arrival of Great Auks there each September and noted that they remained, albeit some distance offshore, until about January. The timing of their arrival in Greenland waters suggests that these might have been immature birds, and indeed an examination of skeletal material from archaeological sites in Greenland supports this idea (Salomonsen 1950; Meldgaard 1988). As with other members of the family, Great Auks in their first year have much less robust skulls, smaller beaks and generally smaller skeletons than adult birds (Gaston 1984). Dick Brown has speculated that the Great Auks visiting southwest Greenland may well have come from the colonies in Iceland making use of the East Greenland Current, as the extant auks breeding in both Iceland and further north still do (Brown 1985). The auks from many areas routinely use ocean currents to aid their migration, not least because of their flightless condition during the annual moult which occurs soon after breeding (Fig. 16).

The Great Auks which bred on Funk Island are thought to have spent their winters either on the Grand Banks, off Newfoundland, or further south. In the late 1770s an old hunter told John Audubon that he remembered Great Auks being common in Massachusetts Bay during the winter (see Steenstrup 1868), and archaeological remains of Great Auks have been found even further south in Florida (Burness and Montevecchi 1992).

Great Auks must always have been vulnerable to man. The combination of being large, tasty, flightless and breeding in large numbers on low-lying islands left them extremely susceptible to human predation. By the 1700s, when written historical records started to become regular, Great Auks were confined to a handful of colonies, of which Funk Island was the largest. They had probably already been exterminated at most other breeding sites. For many years before any Europeans visited Funk Island, the Beothuk Indians had made the treacherous journey by canoe to kill and eat Great Auks. Beothuk arrow heads and a paddle were found on Funk Island in the area now known as Indian Gulch at the east end of the island. Both Joseph Banks and George Cartwright (see Chapter 5) described how the Beothuks visited Funk Island once or twice each year to take the eggs of Great Auks and other seabirds. The dried yolks were an important part of the Indians' diet, being used in soups and to make a type of sausage (Montevecchi and Tuck 1987). The last record of the Beothuks attempting to visit Funk Island was in 1792: they were seen approaching in their canoes by a group of five Europeans who had arrived earlier to collect Great Auk eggs for themselves. The Europeans opened fire on them and the Indians headed back towards the mainland.

The first European to sight 'The Funks' was probably Caspar Corte-Real as he made his crossing from Portugal to Newfoundland in 1501 and what is almost certainly Funk Island appears on a chart made just 2 years later and referred to as Y.-dos-Aves (Isle of Birds), for obvious reasons (Harrisse 1900). By the 1520s crews from several European countries were fishing regularly in Newfoundland waters and it seems likely that Great Auks and other seabirds on offshore islands were providing them with a ready supply of fresh meat. The crossing from Europe to Newfoundland was considerably more time-consuming and difficult then than it is now and Jacques Cartier took 3 weeks to travel from St Malo to St Catherine (now Catalina), Newfoundland in 1534. He reached Funk Island on 21 May and his is the first really detailed description of how mariners viewed Great Auks.

> We . . . sailed . . . as far as the isle of Birds, which island was completely surrounded and encompassed by a cordon of loose ice, split up into cakes. In spite of this belt [of ice] our two long boats were sent off to the island to procure some of the birds, whose numbers are so great as to be incredible, unless one has seen them; for although the island is about a league in circumference, it is so exceedingly full of birds that one would think that they had been stowed there . . . Some of these birds are as large as geese, being black and white with a beak like a crow's. They are always in the water, not being able to fly in the air, inasmuch as they have only small

Indian Gulch on Funk Island.

Gannet (with Herring Gull behind): about 2500 pairs of Gannets breed on Funk Island.

> wings about the size of half one's hand, with which however they move as quickly along the water as the other birds fly through the air. And these birds are so fat it is marvellous . . . In less than halfe an houre we filled two boats full of them, as if they have bene stones, so that besides them which we did eat fresh, every ship did powder and salt five or sixe barrels full of them.

Cartier's account (in Biggar 1924) makes it clear that there were also Common Guillemots and Gannets breeding on Funk Island at this time.

Remarkably, the first tourists arrived on Funk Island just 2 years later. In 1536 a London leather merchant, Richard Hore, took 30 gentlemen in two ships, the *Trinity* and *Minion*, to Newfoundland. As part of their package they stopped off at Funk Island, referred to then as the Island of Penguins, where they found the island 'full of great foules white and gray, as bigge as geese, and . . . infinite numbers of their eggs'. They subsequently killed some and pronounced them 'good and nourishing meat'. Unfortunately, the rest of their trip was less successful, for back on the Newfoundland coast they ran out of provisions and some of the tourists starved to death, while others were cannibalized.

The account of this venture did little to encourage further tourism to the

New World, but the cod fishery in Newfoundland waters continued to expand. Increasing numbers of Portuguese, English, French and Spanish boats spent the summer in this region and the persecution of Great Auks and other seabirds continued unabated. Indeed, in the 1580s the French fleet no longer bothered to bring fresh stores with them, but relied instead upon Funk Island's Great Auks, together with the other birds, fish and marine mammals they could kill (Parkhurst 1578, in Lucas 1890). Great Auks must have provided a very welcome, easily acquired source of fresh meat for the hungry crews of fishing boats. Richard Whitbourne (1622) describes how Great Auks were driven 'hundreds at a time' down gangplanks into waiting boats, adding that 'God made the innocence of so poor a creature to become such an admirable instrument for the sustenation of man'. In fact Whitbourne probably cribbed the first part of this account from Anthonie Parkhurst (1578, in Lucas 1890), which in itself may have been something of an exaggeration—I cannot imagine any auk being persuaded to walk down a gangplank into a boat.

The numbers of Great Auks in Newfoundland waters must once have been considerable since they were regarded as valuable navigational aids. An account written in 1579 states: 'Know that when you approach close to land, about 100 leagues, you will find large black birds that are not able to fly and the Basques call them Dukes [Great Auks] . . .' (Montevecchi and Tuck 1987). Similar information was given in *The English Pilot* published in 1706, which stated that the presence of Great Auks would let mariners know that they had reached the Grand Banks. There is no way of telling how many Great Auks there were on Funk Island at this time, but it is possible to guess. If the Great Auk occupied that part of Funk Island now utilized by Common Guillemots, but, being larger, bred at just one quarter of the density, then there may have been 100,000 pairs of Great Auks breeding there.

Exploitation continued through the 16th, 17th and 18th centuries, initially for meat, but later for feathers, fat and for use as bait. The birds were herded into stone pounds, the remains of which still exist on Funk Island, where they were killed. The birds were then scalded in cauldrons of boiling water to facilitate the process of removing their feathers. Because there was no wood on Funk Island the fatty bodies of the birds themselves were sometimes used to fuel the fires. The discarded carcasses formed the basis of the soil in the grassy area where Puffins now breed. Indeed, the original scientific monograph on the Great Auk, written by Richard Owen (Darwin's adversary) in 1865, was based on one of three skinless mummies that had been found buried under soil and ice on Funk Island 2 years previously (Lucas 1890).

George Cartwright (see Chapter 5) had the foresight to see where the unregulated exploitation of Great Auks was leading. While he was moored in Shoal Cove on Fogo Island, northeastern Newfoundland in early July 1785 he wrote:

> A boat came in from Funk Island laden with birds, chiefly penguins [Great Auks] . . . But it has been customary of late years, for several crews of men to live all the summer on that island, for the sole purpose of killing birds for the sake of their feathers, the destruction which they have made is incredible. If a stop is not soon put to that practice, the whole breed will be diminished to almost nothing,

Rock cairns on Funk Island, made from the stones that were used to build the corrals into which the Great Auks were driven and killed. (Photo: D. N. Nettleship).

> particularly the penguins: for this is now the only island they have left to breed upon . . .'.

It is unlikely that Cartwright had the Great Auk's well-being in mind, since much of his income had been based on hunting and trapping. Rather, it seems likely that this was a plea for a common sense attitude to their exploitation.

Following Cartwright's recommendations, the prevention of egging and the killing of seabirds on Funk Island was enforced, making use of the 'Act against destruction of Wildfowl', passed in 1533. Several men from Greenspond were convicted by Chief Justice Reeves of taking Great Auk eggs from Funk Island in 1793 and sentenced to a public flogging. However, one man, named Clark, was able to prove that he took the eggs only to prevent his wife and children from starving. This obviously appealed to Reeves' better judgement, and Clark enjoyed the privilege of being flogged in private. There are, no doubt, many who today would feel that flogging, either public or private, would be appropriate treatment for those who flout the laws designed to protect wildlife.

* * *

What do we know about the life of the Great Auk, and what can we deduce from comparisons with living auks? We know from contemporary accounts, that like its closest relatives the two guillemots and the Razorbill, the Great Auk was a colonial species, and that like them it laid a single egg. We do not know whether Great Auks spaced themselves out at the colony, as Razorbills

do, or whether they were packed together like Common Guillemots (see below). Judging from the numbers taken from Funk Island I am inclined to believe they might have bred at very high density. To help me form a visual image of Great Auks at their breeding colony I started to think about the posture they might adopt while incubating. The two guillemot species incubate in a more or less upright posture, packed shoulder to shoulder, whereas the Razorbill adopts a much more prone and isolated position while incubating. These positions are determined to some extent by the shape and size of the egg and the location of the adult's brood patch. Common and Brünnich's Guillemot produce relatively pointed eggs and have a single, centrally placed brood patch, while the Razorbill has a more conventionally shaped egg and a brood patch under each wing. Martin Martin (1698) records that the Great Auk had a single brood patch like the two guillemot species. As far as the shape of the Great Auk's egg is concerned, simply from looking at a few eggs it is apparent that their eggs were rather more pointed than Razorbill eggs. However, subjective impressions like this do not carry much weight in science, so I tried to make a more quantitative comparison of the shape of guillemot, Razorbill and Great Auk eggs. Fortunately, Tompkinson and Tompkinson (1966) have published a collection of photographs of just about every Great Auk egg in existence, 75 in total (five other eggs known to exist were not photographed), and I was able to measure the proportions of their eggs from these photographs. The index of shape that I used was the ratio of the maximum width of the egg to the distance between the point of maximum width and the blunt end of the egg. Very pointed eggs (referred to as being pyriform) had ratios of greater than 2, whereas more conventionally shaped eggs had values closer to 2. As expected, Common and Brünnich's Guillemot had the highest values (both about 2.60) and the Razorbill the lowest (2.41); the Great Auk was intermediate, but closer to the guillemots' with a value of 2.58.

Popular accounts of guillemot biology continue to perpetuate the myth that their extremely pointed, or pyriform egg has evolved to minimize the chances of it rolling off the breeding ledge. The idea is that if the egg is knocked then its pointed shape allows it to spin like a top, rather than rolling away. This conjures up all sorts of ridiculous images for me: does the egg spin on its pointed end, as one would imagine a top doing? The spinning top idea obviously has an intuitive appeal because it is rarely challenged and continues to appear in bird books including those for children and adults. However, you have only got to watch a Common Guillemot accidentally knock its egg and watch it roll into evolutionary oblivion, to realize that the spinning top idea is nonsense. The Yorkshire eggers also knew it was nonsense almost a century ago (Wade 1907). The crux of the problem is that although a Common Guillemot egg will often roll in a circle, the circle it describes is far larger than most of the ledges on which Common Guillemots breed. Paul Ingold (1980) spent several breeding seasons rolling Common Guillemot and Razorbill eggs around ledges and came to the same conclusion. The reason why guillemot and Great Auk eggs are such an unusual shape remains a mystery.

Breeding in the open probably forces Common Guillemots to incubate in an upright posture. They do this with the egg tucked between their legs with its blunt end facing forwards. Given that the Great Auk also had a single, centrally placed brood patch and a pointed egg, it seems likely that like the Common Guillemot it too incubated in an upright position (Fig. 9). This in turn lends support to the idea that Great Auks may also have bred in very close proximity to each other. If true, then we might also have expected them to have highly variable eggs which would enable them to distinguish their own from that of their neighbours, as occurs in the Common Guillemot (see Chapter 6). The available evidence confirms this: Martin Martin (1698) described Great Auk eggs as 'variously spotted, black, green and dark'. It is

Fig. 9 *Incubation posture of three auks in relation to the shape of their egg. The Razorbill has a relatively well-rounded egg, lateral brood patches and a prone incubation posture. Note that the Common Guillemot and Great Auk, which have relatively pointed eggs, and a single, central brood patch have a much more upright incubation posture.*

also clear from the black-and-white photographs in the Tompkinsons' book, and from my own examination of the Great Auk eggs in the Cambridge University Zoology Museum, that the markings are as diverse as those on guillemot eggs. There are two colour plates in Seebohm (1896), which further confirm the variability of Great Auk eggs.

* * *

One of the most intriguing questions about the Great Auk's lifestyle concerns the way its chick developed. The feature that makes the whole auk family absolutely unique among birds is the variation in the way the young birds grow up. At one extreme, the Puffins' chick requires as long as 7 weeks from hatching to reach a state where it can leave the colony. It does so at almost adult size, capable of flight and quite independently of its parents. At the other extreme are the most precocious of auks: the murrelets (Ancient, Japanese, Craveri's and Xantu's), all of which occur only in the North Pacific Ocean. These relatively small auks (adult Craveri's weigh about 150 g, and Ancients about 200 g) lay two eggs and their chicks remain at the colony for just 48 hours before departing. By the time these chicks leave the colony they can control their own body temperature, and although they are flightless they can swim and dive very efficiently. An intermediate pattern of development occurs in the two guillemots and Razorbill: they spend about 21 and 17 days, respectively, at the colony and depart weighing less than a quarter of their adult weight and completely flightless. Their father accompanies them as they leave and he then cares for them for several weeks at sea. These three patterns are referred to as semi-precocial, precocial and intermediate, respectively. As Gaston (1992) has pointed out, given that the auk's closest relatives, the gulls, all have semi-precocial young, it seems likely that the intermediate and precocial fledging strategy evolved from a semi-precocial one. The question is, however, which of these patterns did the Great Auk follow?

To answer this we can make some predictions from what we know about the living auks and then try to test these ideas using the information that is available. One could argue that this is a waste of time: Great Auks are so similar in their general appearance to Razorbills that, surely it is safe to assume that like them, their chicks followed the 'intermediate' development pattern, and fledged at about 17 days old. However, if we make that assumption we ignore the fact that the Great Auk's unique flightlessness may have resulted in natural selection operating on it in a completely different way from other auks. Obviously, none of this speculation would be necessary if any of those early observers had left written descriptions of Great Auk chicks or recorded the length of time the chick was present at the colony.

It seems very unlikely that Great Auk chicks followed the Puffins' mode of development, spending several weeks at the colony and fledging when close to adult size. Could they, then, have followed a more precocial strategy, like the chicks of the murrelets?

If the Great Auk had had precocial young we might predict that (i) there would be few records of young birds being seen at the colony, (ii) their breeding season would have been short compared with other auks with semi-precocial or intermediate young, and (iii) they produced a relatively large egg.

First, there are no written accounts of chicks being seen at the breeding colonies, suggesting that young birds were rarely encountered there. Nor are there any remains of Great Auk chicks in any museums (Grieve 1885). Indeed, there are few written records of Great Auk chicks anywhere. Newton's (1861) account is second hand and slightly tangential. He describes a visit to Geirfuglasker (literally, gare-fowl skerry), off southwest Iceland, in early August 1808, by John Gilpin and the crew of his 22-gun privateer *Salamine*, who 'remained a whole day, killing many birds and treading down their eggs and young'. There is a similarly vague and second-hand account by Audubon, who says that fisherman killed the young for bait (Lucas 1890: 495). It is important to bear in mind that one reason for the lack of reference to young Great Auks might be accounted for 'by the fact that after the merciless slaughter of the Auks had fairly commenced, few, if any, eggs were allowed to hatch' (Lucas 1890).

Second, there is some evidence that the Great Auk's breeding season was relatively short. Judging from Martin Martin's account, written in 1698, Great Auks made their appearance at St Kilda off the west coast of Scotland (Fig. 8) at the beginning of May and had left again by mid-June. Because of the change to the calendar, made in 1752, we need to add 14 days to these dates, so they arrived in mid-May and left, just 7 weeks later, in late June. If correct, this implies a relatively short breeding season compared with other North Atlantic auks (Table 4.1). The time that seabirds remain at their breeding colony is determined largely by the duration of the two main phases of the breeding cycle: the incubation period, and the chick-rearing period. Nothing is known about the incubation period of the Great Auk, but it is possible to estimate this using information from other species, and from this we can also estimate how long the young Great Auk might have remained at the colony. Considering the extant auks (and indeed birds in general), the larger a species'

Table 4.1. Duration of the breeding cycle of the Great Auk and other North Atlantic auks.

Species	Incubation period (days)	Chick-rearing period (days)	Total duration (days)
Great Auk	43[1]	?	49[2]
Razorbill	35	17	53
Common Guillemot	32	21	53
Brünnich's Guillemot	32	21	53
Little Auk	29	27	56
Black Guillemot	34	28	62
Atlantic Puffin	40	38	78

All data from Birkhead & Harris (1985), except [1] which is from Harris & Birkhead (1985) and [2] which is from Martin (1698—see text).

eggs the longer is the period of incubation and this indicates that Great Auk eggs probably required 43 days (about 6 weeks) of incubation (Bengston 1984; Harris and Birkhead 1985). If we then assume that Martin (1698) was correct about the Great Auk's breeding season lasting 7 weeks, the young Great Auk must have left the colony when only 5 days old.

In his monograph on the species Grieve (1885) states: 'it is generally supposed that shortly after the young one was hatched it betook itself to the sea, as when it came from the shell it was fitted for swimming and diving'. There is further support for the idea that Great Auk chicks went to sea at an early age since Nicholas Denys (1672) recorded that Great Auks were abundant on the Grand Banks where they were seen to carry their young on their backs. It seems likely that if a Great Auk chick was small enough to be carried at sea on an adult's back it must have been small and hence precocial, when it left the colony. Otto Fabricius, writing in the late 1700s (unpublished manuscript quoted in Meldgaard 1988), mentions occasionally encountering down-covered Great Auk chicks on the sea during August off the west coast of Greenland. Grieve (1885) thought that Fabricius mistook some other species for Great Auk chicks and subsequent authors have followed Grieve in ignoring this information, mainly because Great Auks were thought not to breed in west Greenland. However, as Meldgaard (1988) points out, Fabricius was so meticulous and accurate in his other natural history observations, there is no reason why these particular ones should be disregarded. The significant part of Fabricius' observation is that when he dissected the chicks he found nothing but some fragments of rocky shore-dwelling plants in their stomachs indicating that they had only recently fledged and had not been fed by their parents.

Third, we know that the Great Auk laid a large egg, but was it unusually large for the size of bird, and what might we infer from the answer? One way to explore this is to consider the weight of an average Great Auk egg relative to the size of the bird itself and then compare this with the other auks. Like most other auks, the Great Auk laid a single egg, similar in shape to that of the Common Guillemot. Although no-one ever weighed one, the weight of a fresh Great Auk egg can be estimated from its linear measurements and a comparison with Common Guillemot eggs. An index of the volume of any species' egg can be obtained by multiplying the length by the breadth squared ($l \times b^2$), and this index is closely correlated with the fresh weight of an egg. The eggs have to be fresh since those of most species lose about 15% of their initial weight through evaporation during the course of incubation. By plotting the relationship between volume index and weight for Common Guillemot eggs and then extending the volume index until we reach that of the Great Auk we arrive at an estimate of 327 g for the weight of a Great Auk egg. This is four or five times that of a chicken egg, or about 6.5% of the weight of an adult Great Auk.

Next, we can look at the relationship between egg weight and body weight across all auk species. When we do this it is hardly surprising to find that bigger auks lay bigger eggs (Fig. 10). Since this is a fairly tight relationship (the points are tightly clustered around the straight line), we can use it to

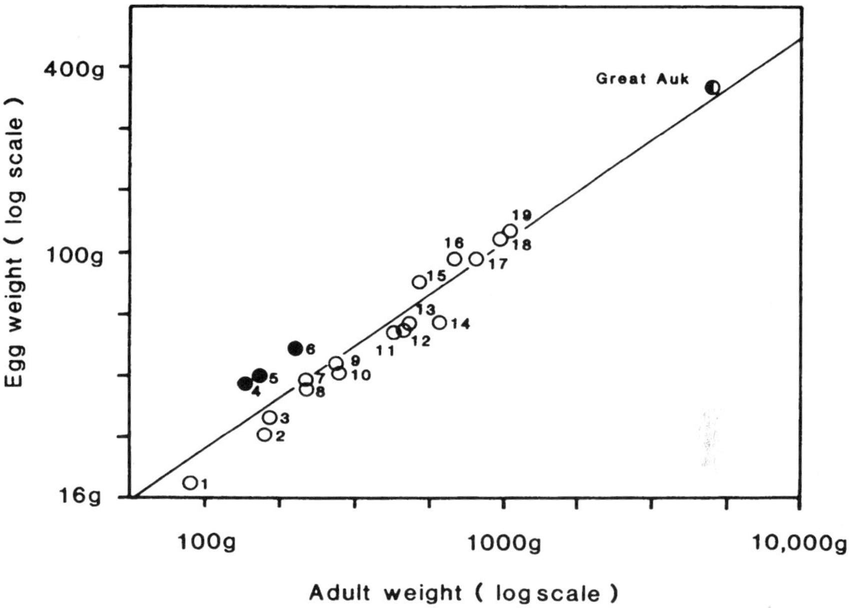

Fig. 10 *Relationship between body weight and egg weight in the auks. The plotted regression line is calculated using the information only from the extant species (regression equation:* $y = 0.129 + 0.716x$*) and is highly significant (*$r = 0.961$, $p<0.001$*). Note that the point for the Great Auk falls almost directly on the line, indicating that its egg size is almost exactly as we would have predicted from the extant auks. Note also that the eggs of the three species with precocial young (filled symbols) fall well above the line and hence are relatively large. 1. Little Auk, 2. Cassin's Auklet, 3. Least Auklet, 4. Craveri's Murrelet, 5. Xantu's Murrelet, 6. Ancient Murrelet, 7. Marbled Murrelet, 8. Kittlitz's Murrelet, 9. Crested Auklet, 10. Parakeet Auklet, 11. Black Guillemot, 12. Pigeon Guillemot, 13. Atlantic Puffin, 14. Horned Puffin, 15. Rhinoceros Auklet, 16. Razorbill, 17. Tufted Puffin, 18. Brünnich's Guillemot, 19. Common Guillemot.*

determine whether Great Auks produced relatively large eggs or not. It is clear from Fig. 10 that the data points for the precocial murrelets lie well above the fitted regression line, indicating that for their body weight they produce relatively large eggs. The points for the three intermediate species lie very close to or slightly above the line, indicating that they too produce slightly larger eggs than expected for their body weight. In contrast, semi-precocial species produce relatively small eggs, with the data points for 10 of the 13 species falling below the line. To be able to make our comparison with the Great Auk we need to know its weight. Remarkably, despite the thousands of Great Auks that were killed there is only a single record of one being weighed: Feilden (1872) noted that one weighed 9 Danish pounds (4500 g). Later, Bédard (1969) using the Great Auk's dimensions estimated that the average weight was about 5000 g and independently, Livezey (1988) came up with an almost identical value of 4999 g. Using a value of 5000 g in Fig. 10 shows that Great Auk's egg falls directly on the line, indicating that it produced an egg almost exactly the size predicted from its body weight: neither larger nor smaller. On the basis of this evidence alone we would be forced to conclude that the Great Auk chick followed a similar course of development no different from the 'intermediate' species, the guillemots and Razorbill.

However, although the Great Auk did not produce an especially large egg for its body size, we cannot entirely exclude the possibility that it produced a precocial chick. After all, Ancient Murrelet chicks weigh just 27 g when they go to sea at 2 days of age (Gaston 1992), and judging from the dimensions of their eggs Great Auk chicks would have weighed 236 g at hatching (Harris and Birkhead 1985). Therefore, in terms of absolute body weight, there is no reason why Great Auk chicks could not have been precocial.

Taken together then, the information from our three predictions is consistent with the idea that Great Auk chicks were indeed precocial. However, there is one further piece of evidence that makes this interpretation compelling. Although adult Great Auks were strong swimmers (Grieve 1885), their flightlessness must have meant that the distance over which they foraged from their breeding colony was more limited than the other auks, some of which, like Brünnich's Guillemot, may travel over 200 km on a round trip. Olsen *et al.* (1979) suggested that Great Auks may have foraged within just a few kilometres of their breeding colony (but see Brown 1985). As some compensation, their large body size will have given them the potential to make deep and lengthy dives in pursuit of their fish prey (see Chapter 6). Notwithstanding, for a flightless bird the energetic costs of ferrying food back to a chick at the colony would have been prohibitively high (see Prince and Harris 1988) especially as one parent would have had to stay behind and guard the chick. It therefore seems likely that they would have avoided this problem by taking their chick to sea at a relatively early age, before its demands for food outstripped the adults' ability to feed it (see Fig. 16, Chapter 6).

* * *

What did Great Auks eat? The only direct observations were those of Fabricius who dissected Great Auks killed in western Greenland in the 1770s. He noted that they contained two species of fish: the shorthorn sculpin and the lumpsucker (O. Fabricius cited in Salomonsen 1950). Other biologists have tried to reconstruct the diet of the Great Auk using indirect, but often ingenious methods. In 1887 Frederic A. Lucas was part of an expedition which visited Funk Island to collect Great Auk remains. They found huge numbers of individual Great Auk bones, but none of the entire mummified carcasses obtained by an expedition in 1863, which was ostensibly seeking guano. The skeletal material obtained by Lucas (1890) was lodged in the soil that was actually formed from the corpses of generations of other Great Auks massacred in previous years. One wooden crate that Lucas had filled with Great Auk bones remained much as he had left it for almost a century in the basement of the National Museum of Natural History at the Smithsonian Institution in

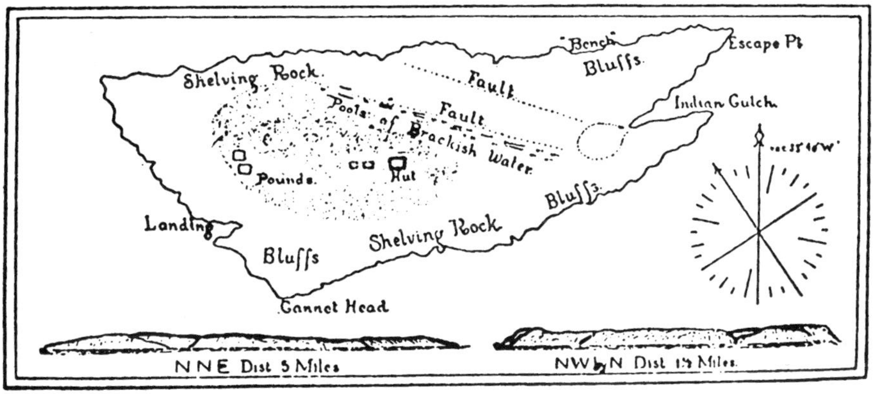

F. Lucas' map (1890) of Funk Island. (From Montevecchi and Tuck 1987).

Washington D.C.. In the late 1970s Storrs Olson and his colleagues examined the bones in this box, focusing on the soil adhering to the bones in search of fish fragments. They found fish scales, some vertebrae and a few other bits, including a spine and fin-ray, but no otoliths (ear bones), which were probably destroyed in the acidic soil. Olson *et al.* (1979) were then able to identify and estimate the size of some of the fish to which these remains belonged. Innovative as this method was of assessing the Great Auk's diet, it was not totally convincing. Bradstreet and Brown (1985) pointed out that several of the fish species they identified (such as the menhaden *Brevoortia* and a perch or bass *Morone*) were unlikely to have ever lived anywhere near Funk Island, and their remains may actually have been transported to Funk Island by man himself. For example, the crate which Lucas used to transport his Great Auk material might have previously been used to transport menhadens. Storrs Olson pointed out that, judging from the label it carried, the crate had previously been used to carry Stejnegner's specimens of Steller's Sea Cow from Bering Island, and it was unlikely to have held menhadens. (Steller's Sea Cow was exterminated by man in the late 1700s). Another possible explanation was that the scales of these fish were already on the old fishing nets that were used to capture and herd the Great Auks together before they were killed (Bradstreet and Brown 1985).

In Sherlock Holmes fashion, Keith Hobson and Bill Montevecchi (1991) have used another an ingenious and high-powered technique to find out more about the Great Auk's diet. They cleverly exploited the fact that tissues, such as collagen (the fibrous, proteinaceous material in bone) contain chemicals which reflect their previous owner's diet. Their results indicated that Great Auks ate both crustaceans and fish. Their most interesting result however, came from a single bone of a young Great Auk. The analysis of this bone prompted Hobson and Montevecchi to suggest that the young Great Auk might have been fed largely on plankton by its parents while at the colony.

This is a remarkable result and could open up a whole new scenario about the biology of Great Auks. However, before jumping to conclusions we should remember that this result is based on only a single bone, which is only thought to be that of a young Great Auk. Moreover, it is entirely possible that young Great Auks may have fed *themselves* on plankton as soon as they departed from the colony.

* * *

The Great Auks became extinct on Funk Island in about 1800, and the space they once occupied is now taken over by one of the largest colonies of Common Guillemots anywhere in the world. As we landed at The Bench on the north side of the island, the large number of Common Guillemots in the sky and on the water provided some indication of what was awaiting us over the top of the low cliff. However, neither these nor the photographs I had seen of Funk Island prepared me for the carpet of birds that we then saw. My mind grappled unsuccessfully with the curious juxtaposition of birds and topography as I tried to estimate their numbers. Because there were no vantage points, one obtained a curious perspective on this huge aggregation of birds. Most seabird colonies I had visited had been arranged vertically, this one was horizontal and it made it particularly difficult to get any feel for the numbers of birds.

Along the central axis of the island, there were three separate areas each containing countless (meaning uncountable) numbers of Common Guillemots. The density at which they were breeding was as remarkable as their sheer numbers. And it was their density that we were to try and measure in order to estimate the total population size.

The plan was as follows. On this visit we would count the number of pairs of Common Guillemots in a sample of 1 m^2 square quadrats throughout the island. The following day, aerial photographs of the breeding areas would be taken so that the size of the total area occupied by the Common Guillemots on Funk Island could be measured. We would then combine the two pieces of information to arrive at an estimate of the total number of breeding pairs. All this sounded simple enough, but once on the ground, at the edge of the colony, the problem took on different proportions.

We started in Indian Gulch where the Common Guillemots bred on an absolutely level, uniform surface of fine gravel. On closer inspection the gravel was not stationary: it heaved gently with a mass of ticks. I tucked my trouser bottoms into my socks and tried to ignore them. By approaching groups of birds extremely carefully we were able to persuade the incubating birds to get up off their eggs and move away so that they did not leave the island in a panic. We could then count their eggs, and allow the birds to return as soon as possible after our intrusion. The number of eggs in these areas was remarkable: virtually every bird was incubating an egg. Cautiously stepping between the eggs we laid out our rope quadrat, knotted at 1 m intervals into a square. I started counting and ended up with a number I could not believe. Keeping track of a large number of eggs was tricky and I was convinced I had made a mistake. I re-counted, and got the same answer: 76 eggs in 1 m^2! I thought

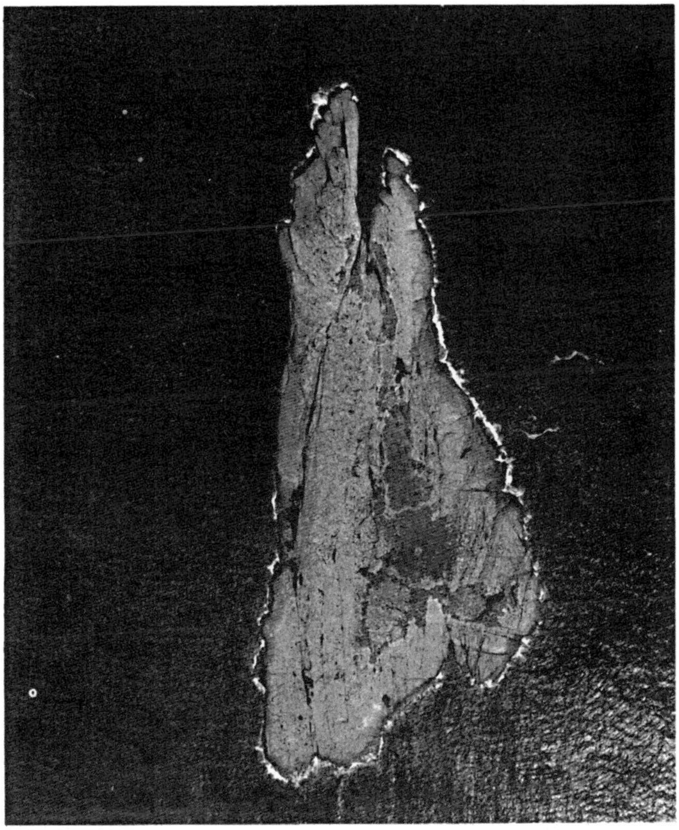

Aerial view of Funk Island, looking from west to east, with Indian Gulch at the far end. The darker circle in the centre of the island is a grassy area occupied by Puffins. The darker areas before and beyond it are Common Guillemots, and the Gannet colony lies to the bottom left of the Puffins. (Photo: D. N. Nettleship).

then that maybe not all the eggs were currently being incubated, so I felt each one in turn to check whether it was warm. They all were.

We went on counting quadrats of eggs in different parts of the island and on different habitats, some on gravel others on rock. We did not find any others as dense as that first one, but some came close. The consistency of the gravel areas yielded the highest densities because all of this habitat was equally suitable for breeding. The average density of incubated eggs here was 39.14 (range 17 to 76 in 14 quadrats). As one might expect, the habitat in rocky areas was much less uniform, and densities were lower, averaging 19.5 in each square metre. Overall, the average density was about 20 eggs in each square metre. Multiplying this by the area occupied by Common Guillemots (determined from the aerial photographs) gave an estimated population of about 400,000 pairs.

Common Guillemots breeding on gravel in Indian Gulch on Funk Island. The group of birds in foreground are some of those whose eggs we counted in the rope quadrat.

High density nesting in Common Guillemots at Funk Island. There are 76 eggs within the 1 m² rope quadrat.

This kind of approach to censusing Common Guillemots left a lot to be desired since it was extremely unpleasant to have to disturb the birds. Admittedly our disturbance was short-lived and within an hour most birds were back on their eggs. Also, by being careful we affected only a tiny proportion of the birds. One always has to weigh up the costs and benefits of such invasive techniques. Our justification was that by trying to standardize our census techniques we, or others, would be in a better position in the future to assess whether a change in the population size had occurred at this vast seabird metropolis. All previous estimates of the Funk Island Common Guillemots had been made by guessing (Montevecchi and Tuck 1987).

* * *

The last Great Auks on Funk Island were killed in about 1800, just as George Cartwright had predicted. Frederick Lucas described his visit to Funk Island in 1863 thus: 'Here to day the bones of myriads of Garefowl lie buried in the shallow soil formed above their moldered bodies, and here, in this vast Alcine cemetery, are thickly scattered slabs of weathered granite, like so many crumbling tombstones marking the resting places of the departed Auks' (Lucas 1890).

In Britain, at St Kilda, superstition resulted in the last Great Auk being killed as a tempest-conjuring witch in 1843 (Ley 1935). The increasing scarcity of Great Auks in the early 1800s meant an accelerating decline as the market value of skins and eggs rocketed. As a result, visit after visit was made to the last known breeding sites off southwest Iceland during the 1830s and '40s and it is thought that virtually all eggs currently in existence came from this area. Following their demise Great Auk artefacts continued to attract high prices. As one 19th century British newspaper wrote anticipating the sale of a skin: 'This relic of the sea is a stuffed Great Auk, one of the seventy-nine which still remain to stir spectacled old men to frenzy and inspire awe in the bosoms of those to whom the unattainable is the pinnacle of desire' (from Ley 1935).

In the sense that the last known individuals were deliberately killed, man exterminated the Great Auk. However, it is worth noting that its fate may have been sealed long before man became a serious threat. The Great Auk may never have been particularly abundant. Its flightlessness meant that it would be restricted to a small number of localities, where it could hop directly from the sea to its breeding sites, and where food was abundant relatively close by. Since it bred at just a few sites (see also Chapter 3) it will have been especially vulnerable to chance extinction, such as the appearance of novel predators (see Chapter 9), and climatic change. For example, the 'Little Ice Age' started in the 13th century, and certainly resulted in the extinction of some human populations (Grove 1988). The Icelandic sagas provide detailed confirmation of this cold period between 1200 and 1600 and indicate the amount of sea-ice around the coast was extreme during this time. Heavy sea-ice would certainly have prevented Great Auks from reaching their colonies and may, over a period of time, have resulted in the extinction of some populations. Bengston (1984) has suggested that environmental changes in the north Atlantic meant Great Auks may have already been well down the road to extinction by about 1500 when man seriously started to exploit them. Not that this is any excuse.

If these events had occurred a century later it is almost certain that, like the California Condor, the last survivors would be eking out their final days in some well-funded zoo.

CHAPTER 5

Labrador

. . . it is not to be called The new Land, but rather stones and wilde cragges, and a place fit for wilde beastes. . . . To be short, I beleeve that this is the land that God allotted to Caine

Jacques Cartier (1534: in Biggar 1924)

Why is it that flights to remote parts of the world always depart at anti-social times? The flight to Goose Bay, Labrador left Halifax, Nova Scotia at 01.30 h and arrived just as day was dawning. I was full of anticipation, thrilled by the romance of Labrador, but Goose Bay was a dump: a severe disappointment.

Lack of sleep, horizontal rain and blustery wind combined to provide a poor welcome to this sodden air base in the centre of Labrador. Moreover, as it got lighter, but no drier, the image of Goose Bay did not improve. This huge, sprawling, almost abandoned air force base, has a feel of decay.

Situated at the western most end of Lake Melville, Goose Bay (Fig. 11) is one of the biggest, and probably the least attractive of Labrador's settlements. It arose in its present form from the USA's need, during World War II, for a refuelling spot for Atlantic crossings. Even after the war ended Goose Bay air base continued to expand in the wake of deteriorating East-West relations and the build up of the 'cold war'. One spin-off of this was the string of DEW (direct early warning) line stations established along the Labrador Coast in 1951. Their construction provided welcome employment for local people, but subsequently left some ugly scars on an otherwise pristine landscape. However, by 1976 the US airforce had all but abandoned Goose Bay, leaving vast numbers of empty and not so empty buildings. One building where we tracked down the office of a helpful wildlife officer also housed several tens of thousands of old mattresses, stacked up on each other like so many bodies—remnants of a busier time for Goose Bay. The airbase retained a few token air crews, including the Royal Air Force. When I was there the RAF's presence certainly enhanced the tranquillity of the place: the air was filled with the rich redolence of aviation fuel as Vulcan bombers looped the loop just a few hundred feet above the ground.

* * *

This was my first visit to Labrador and was actually my first visit to any Arctic seabird colony. In fact, this trip took place in 1978, in the few weeks immediately before I went to Cape Hay (Chapter 2), and hence also pre-dates my visit to Coburg and to Newfoundland (Chapters 3 and 4). As will become apparent, I had a longer, and more fruitful association with Labrador than any of these other places, and for that reason I have tried to keep all the material relating to Labrador together, even though this disrupts an otherwise tidy chronology.

* * *

Remy and I were heading for Cartwright, one of the largest coastal settlements and where we were to meet up with Richard. Once at Cartwright we were to charter a fishing boat to take us out to the Gannet Clusters, a group of six islands some 45 km from the entrance to Sandwich Bay. Very few settlements in Labrador are connected by road and there were only two ways of getting to Cartwright, either by the route Remy and I were taking via Goose Bay, or on the coastal ferry from Lewisporte in Newfoundland, as Richard had done.

We checked in to the Hotel Goose in the inappropriately named Happy Valley, a settlement adjacent to Goose Bay, while we waited for transport to Cartwright to materialize. Everywhere looked such a mess, abandoned cars, paper and plastic garbage everywhere. Why do people that live in remote areas

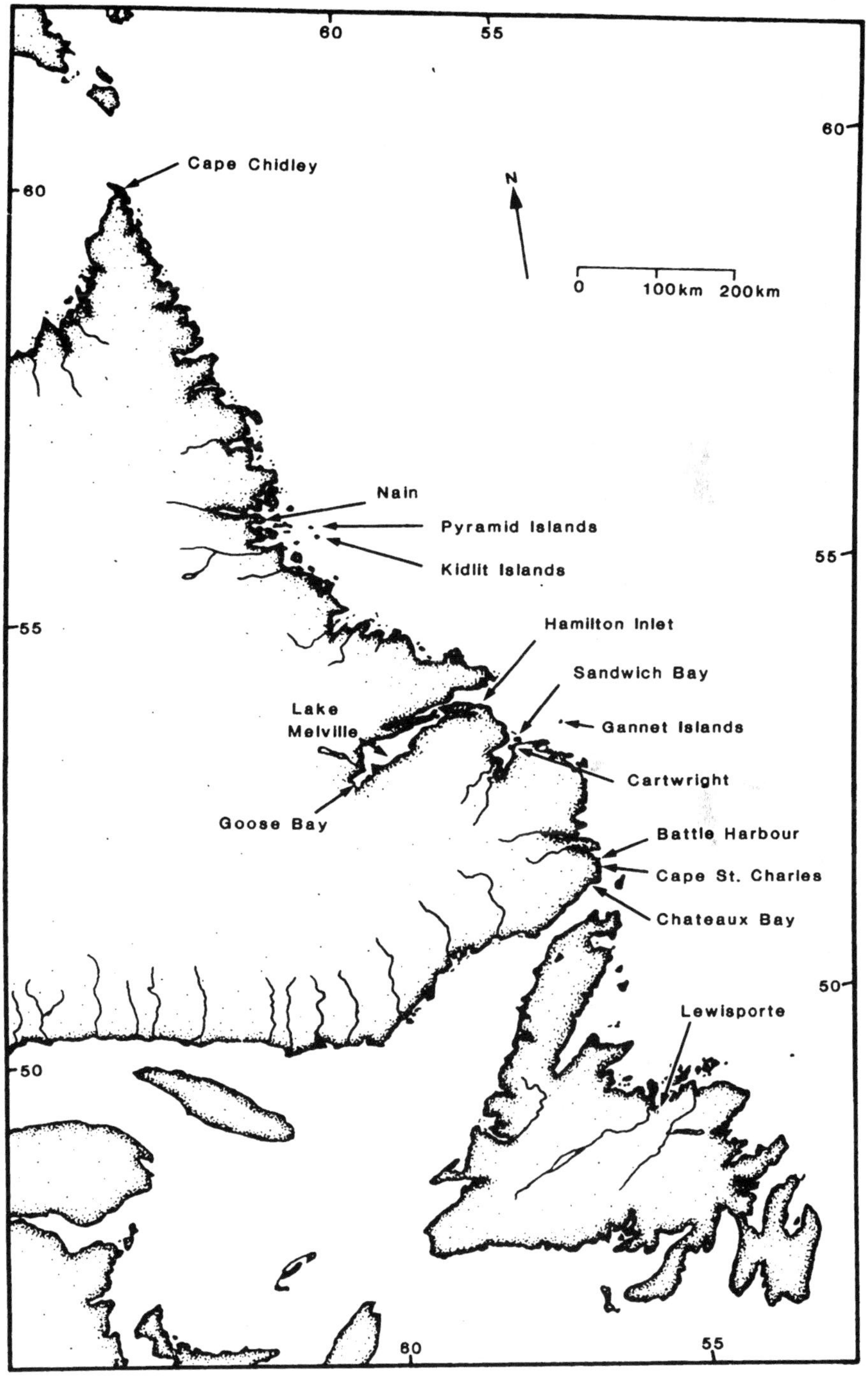

Fig. 11 *Map of Labrador and Newfoundland showing locations mentioned in the text.*

The DEW line station at Cartwright, Labrador, or what was left of it before it was dismantled and replaced in the late 1980s.

treat their immediate environment with such disdain? This was not a unique situation, it is the same everywhere, from Shetland to the Caribbean. I had been warned about Goose Bay, so my expectations had not been especially high, but environmental indifference is depressing. I could not imagine why anyone would want to live here. Apparently, Goose Bay was even worse at certain times of year when the mosquitos and biting blackflies were abundant. I could not help expressing my despair about the place to a taxi driver as he drove us the 10 km from the airport to Happy Valley. I shut up when he told me that he had been born and bred in Goose Bay, and thought it was 'just fine'.

My feelings for Goose Bay changed somewhat over successive visits, and one spring when I arrived at a civilized hour and in the most beautiful weather, I actually felt quite positive about the place. However, on this first occasion all I wanted was to get out of it as soon as possible and on to Cartwright.

After two tedious and virtually birdless days in Happy Valley, a helicopter was arranged to take us to Cartwright. We flew at a few hundred feet over the Mealy Mountains and the beautiful swampy muskeg into the much more attractive settlement of Cartwright. Neither we nor the pilot had much idea where we were going, and he parked the helicopter on some waste ground opposite a house. We called at the house to ask where Richard might be, but were sucked into afternoon tea with the Anglican minister and his wife. In retrospect I felt uneasy about their hospitality; they probably saw us as some kind of bonus: more souls to be saved and a few brownie points scored over the Pentecostal and United ministers.

Tea over, the pilot headed back to the bright lights of Goose Bay, while Remy and I met up with Richard at the Bird's residence, where he had become part of the family after just 2 days lodging. Cartwright was delightful and a complete contrast to Goose Bay. Here the houses were small, single storey and most of them painted in bright colours. The settlement, founded by Captain George Cartwright in 1775, lies in a sheltered cove at the entrance to Sandwich Bay, surrounded by rolling, spruce-covered hills (Fig. 12). Before setting off from England I had pored over Wilfred Grenfell's books on Labrador, most of which were published in the early decades of this century, and was pleased to see how little Cartwright appeared to have changed with the passage of time. In the 1980s Cartwright was the home to about 700 people: there was a Hudson's Bay Store and Fisherman's co-operative, a craft store, fish processing plant, and three churches. Most people's livelihood was tied up in one way or another with the fishery, which is mainly for cod and salmon.

We fixed up some new accommodation for us all and set about trying to find a fishing boat that would take us out to the Gannet Clusters.

* * *

The idea of hiring a fishing boat to get us out to the Gannet Clusters seemed comparatively simple when we were sitting in an office at the Bedford Institute of Oceanography, Nova Scotia. The fact was that because people were busy fishing there were few boats around and none of the skippers we spoke to seemed very keen to take us out to the Gannets. Eventually someone informed

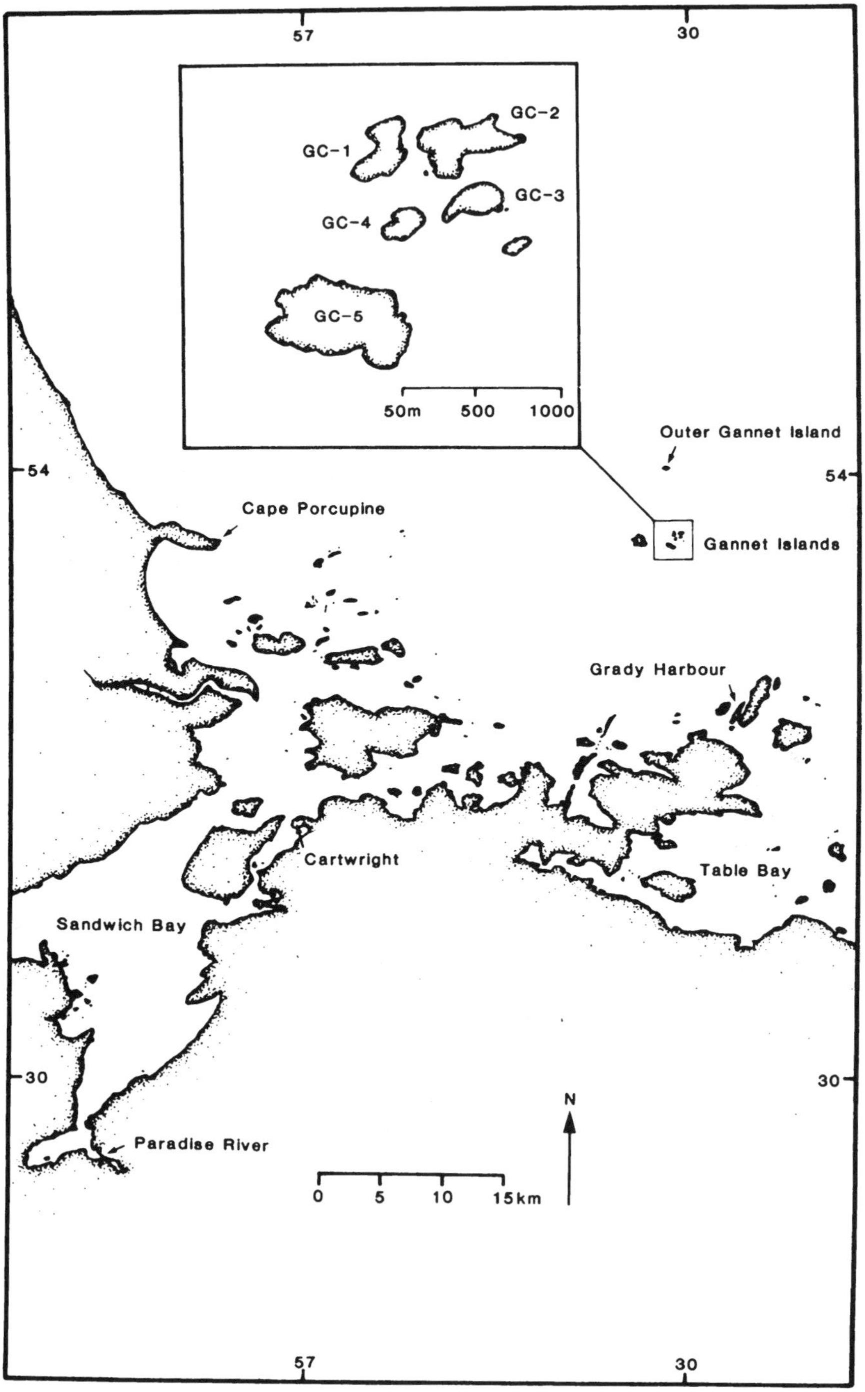

Fig. 12 *Map showing the Gannet Islands and the adjacent mainland.*

us that Ivan Roberts and the *Pauline-Gladys* would be coming in to Cartwright in a couple of days time and might be a good bet for taking us out.

Ivan and his crew were an evil looking bunch. They were unenthusiastic about going out to the Gannets, but finally agreed and told us to be on the boat at 06.00 h the next morning. We were delighted, and in our enthusiasm for getting out there abandoned our comfortable lodgings and slept on the floor of the Forestry hut near the wharf, ready for an early start. We need not have bothered: it was not until after 11.00 h that several hung-over bodies emerged from below decks to say that they were not going anywhere. Another day of sitting around.

'Be here at 04.00 h tomorrow' they said. We agreed, and this time we were off. The crew were in better shape this morning and they provided us with a welcome, cholesterol-rich breakfast in the galley. I was intrigued by the domestic arrangements: Ivan had his own private berth next to the wheelhouse while the crew slept together in four bunks under the prow, which also served as the galley. I was slightly unnerved when I saw that each man slept with a gun in his bunk, and even more apprehensive when, shortly after setting off, the crew brought their guns on deck to let rip at a passing iceberg. Perhaps we should have seen the writing on the wall at that stage. Anyway, it was too late: we were heading for the Gannet Clusters, and it did not seem worth bothering too much about what lay ahead.

The monotony of the 4-hour journey was sporadically broken by the excitement of seeing some icebergs and whales. The icebergs were small but they were the first I had seen, and I was elated. We saw several Minke whales and a single humpback, and literally hundreds of Red-throated Divers (loons) flew overhead as we chugged slowly towards the islands. Auks were surprisingly scarce until we drew close to the Gannet Islands.

I noticed Richard in conversation with the skipper and at a discreet opportunity I suggested that he should not pay Ivan until our return trip. But it was too late, Ivan had insisted on having half the money immediately: I pushed the thought of what this might mean to the back of my mind.

As we came in amongst the islands (Fig. 12) I could see why they were called the Clusters; the four smallest islands formed a tight group and provided a protected mooring. There were birds everywhere: thousand upon thousand of Common Guillemots covered the sheltered, innermost sides of the islands, while the sky was thick with Puffins, there were also Eider Ducks, Harlequins, a few honking Canada Geese, Ravens and on shore we found Spotted Sandpipers and White-crowned Sparrows. My main purpose in coming here was to see Common and Brünnich's Guillemots breeding side by side. The two guillemot species replace each other geographically, the Common Guillemot occupying warmer waters than its Arctic counterpart. In a few locations, including the Gannet Islands, the range of the two species overlap. In the previous survey in 1972 the Common Guillemot was the most abundant seabird at the Gannets, and while Brünnich's Guillemots had been found on Outer Gannet they were not known to breed on the Clusters. I was pleased therefore when, before even setting foot on the islands, I saw a group comprising both species on one of the Gannet Clusters.

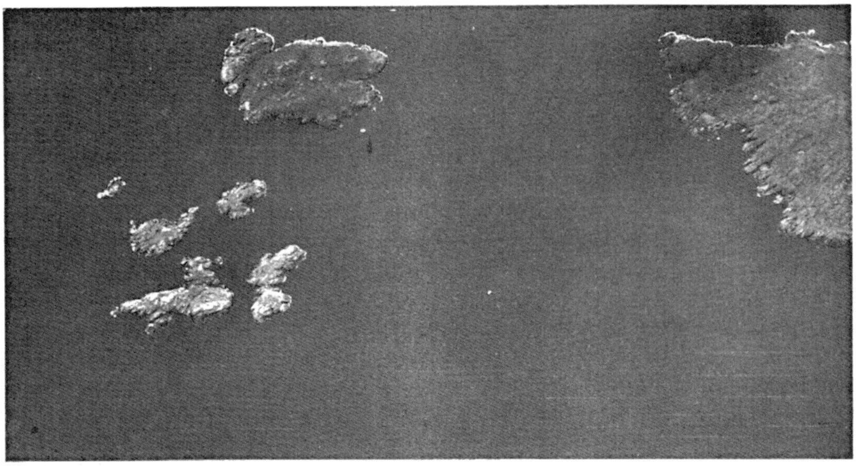

The Gannet Islands from the air. The four islands forming the clusters are (from the top clockwise) GC4 (the smallest), GC1, GC2 and GC3. The larger island above them is GC5, and to the right is part of GC6. The cove where most of our Guillemot observations were made is clearly visible on the right hand side of GC4. Note that this picture was taken looking from north to south, hence the different orientation from Figures 12 and 25.

In less than 1 hour we were ashore with all our equipment including our inflatable Zodiac, and the *Pauline-Gladys* had disappeared. There were so many birds it was difficult to find a spot to erect our tent where we would not disturb them or cover the numerous Puffin burrows. We eventually found a suitable site and sorted out our supplies. As we unpacked, an unfamiliar packet caught my eye and I asked Richard what was inside it. 'Hardtack: a Labrador speciality' he said. But to me the contents looked like white dog biscuits and I told him and Remy that they would not catch me eating it.

In the previous seabird survey of the Gannet Clusters in the 1970s Tony Lock had given the six islands numbers in order to identify them. They were simply GC1–GC6. I now regret that we were not imaginative enough to give the islands proper names. The main cluster of four small islands were GC1, 2, 3 and 4. The next nearest, about 500 m south of GC4, was GC5, and about 3 km from the main cluster lay the biggest island, GC6 (also known locally as Western Gannet). Interestingly, the two largest islands had the fewest seabirds, despite plenty of suitable habitat. All the islands were relatively low-lying; the four islands in the main cluster ranged in height from 25 to 40 m, but GC6 was the highest at 66 m. The islands were rocky, but there were relatively few vertical cliffs. In most areas the rocks had been worn by ice, wind and sea into complex shapes. The vegetation was a mixture of crowberry heath and grass, with a few flowering plants.

Richard and Remy's objective of being at the Gannet Clusters was to census this bewildering mass of seabirds. I had been allowed to join them because it

Common Guillemot on the Gannet Clusters (GC3 in fact) in 1978, showing the irregular breeding habitat. Compare this with the photograph on page 254 (Chapter 9) taken in 1992.

was felt that my experience with Common Guillemots might help. In fact I was rather at a loss. All the other Common Guillemot colonies I had seen had been vertical, with birds lined up on narrow cliff ledges. These colonies were basically horizontal, and the birds bred in dense masses on irregular terrain almost down to sea level. Given the amount of time that was available the idea of counting individual birds was out, there were simply too many and at too high a density.

* * *

How did the Gannet Islands get their name? Given that there is no evidence that Gannets ever bred there, and the nearest ones were about 500 km to the south, at Funk Island, the name seems incongruous. The process I underwent in trying to discover the origin of the islands' name reminded me of an incident that occurred in our department's aquarium in Sheffield. A colleague of mine studies the subtle but often spectacular colour changes in cephalopods: squid, cuttlefish and octopus. One afternoon there was a delivery of live octopus, each in its nervous, retracted shape, about the size of a small loaf of bread. They were duly placed individually in their specially constructed tanks, the lids battened down and the animals left to recover from their journey. The next morning all the animals had disappeared and were assumed to have been stolen. Later, we found them (in various states of vitality) in other parts of the aquarium room. Someone must have released them—at least that is what was thought until, on returning one of the animals to its home, it immediately proceeded to extrude itself through an aeration hole little more than 1 cm in diameter in the tank's lid. The octopus's body, which lacks either an external or internal skeleton, allows it to perform such extraordinary feats.

The process is almost identical to that involved in moving into a new area of research. The intellectual gap through which one has initially to squeeze seems impossibly small, but then, once you have found a key reference, you slip through all too easily and find that there is an even bigger world out there than the one you are used to. My attempts to discover the origin of the islands' name led me into a labyrinth of archives, maps and letter writing, and the (unwanted) potential for years of study. Moreover, it was not like moving into a new area of biology; I rapidly discovered that this type of historical research was rather like starting a jigsaw but without knowing either how many pieces there are, or whether all the pieces were there. Unlike the octopus, having forced my way into this other world, I did what I had to do and returned as quickly as possible to the safety of my scientific container.

One of the first Europeans to see and set foot in Labrador was the Norse explorer, Bjarne Herjulvson in 986. Remarkably, in the present context, his landfall on the Labrador mainland was probably near Sandwich Bay, not far from the Gannet Islands (Tanner 1947). The Norsemen also discovered and settled in Greenland, and in subsequent years sailed from there to Labrador in order to obtain fresh timber. The Norse settlements flourished during the 11th, 12th and part of the 13th centuries, but by the fourteenth century they started to decline and by about 1500 they had disappeared (Nansen 1911). In

1492 however, Columbus set foot in the New World and 5 years later John Cabot discovered Newfoundland. About this time Portugal started to send expeditions to the northwest Atlantic in search of a route to Cathay (i.e. China): Caspar Corte-Real went in 1500 and landed somewhere on the northern Labrador coast. His discoveries together with those of Cabot sparked off considerable interest back home and there were soon large numbers of fishing vessels and explorers from Europe in this region. The rich cod banks of the New World discovered by the Normans, Bretons and Basques in the early 1500s have continued to be exploited virtually ever since (see Chapters 7 and 9).

The first maps of Labrador, produced by the Portuguese and French, start to appear in about 1500s, and although they include a recognizable outline of the entire Labrador coast they provide rather little detail. Similarly, those of the early Moravian missionaries, who played such a central role in the development of Labrador (Whiteley 1964; Lysaght 1971), are also frustratingly imprecise. For almost two centuries the fishermen of several European nations exploited the cod fishery on the Newfoundland and Labrador coast before the British began to realize its economic potential. The turning point was the Treaty of Paris in 1763; once the French had formally ceded Canada to England it was clear that accurate charts of the coast would facilitate England's exploitation and protection of the Newfoundland-Labrador fishery.

Captain James Cook was sent off in HMS *Grenville* in 1766 to conduct a hydrographic survey of the south coast of Newfoundland. During this same summer, Joseph Banks, then 23 years old, set sail on board the fisheries protection vessel HMS *Niger*, to spend 7 weeks at Chateau Bay on the south coast of Labrador. The crew of the *Niger* was there to build a block house, but Banks, who was later to become president of The Royal Society, was able to study the wildlife. Banks' achievements were outstanding, but little recognized, mainly because he never wrote up his observations. The fauna and flora of this part of Canada were virtually unknown in the 18th century and Banks' collection of 234 plant species and 91 species of bird, together with some fish and invertebrates made a substantial contribution to our knowledge of the New World. In the altruistic tradition of medieval scholars Banks sought no recognition for his work in Labrador—he simply allowed others to use his material and publish their own reports (Lysaght 1971).

During Cook's survey of the Newfoundland coast he encountered other British vessels, including HMS *Guernsey*, upon which Michael Lane was the schoolmaster. During the winter of 1766–67 Lane was transferred to the *Grenville* to act as Cook's assistant. The following summer was to be Cook's last one in Newfoundland, for the year after that he set off round the World in HMS *Endeavour* accompanied by Joseph Banks. Michael Lane was left to complete the survey of the Newfoundland and Labrador coasts, which he did over the next seven summers.

It was during August and September 1771 that Lane worked in the Sandwich Bay region and his is the first chart in which the Gannet Islands are named. In an attempt to discover why Lane had so named the Gannet Islands I tracked down the log of his ship, HMS *Grenville*, in the Public Record Office in

London. It was exciting to have this 200-year-old document in my hands, but on turning the fragile pages I was disappointed not to find more detail: entries were sporadic and terse. Even more frustratingly there was no specific mention of the Gannet Islands, and despite having columns for presenting his co-ordinates each day, Lane never did! However, the log makes a number of references to sounding among the Outer Islands, which I am certain must be the Gannet Islands. Lane was at the 'Outer Islands' on 29 and 30 July, and then again on 20 and 21 August and 25 and 26 September. Since his survey covered the rather limited area between Spotted Island and Sandwich Bay there are few other islands in this region that could accurately be described as 'Outer'.

The lack of mention of the Gannet Islands is, however, rather odd since the original chart from that survey, which I located at the British Navy's Hydrographic Department in Somerset, clearly marks the Gannet Islands and Outer Gannet. Also, in his written report of that survey, Michael Lane states 'From Wolf Rock to the Outer Gannet Island the course is N63°W Distance 10 Leagues; these are a cluster of Islands of a moderate height' (Lane 1771).

In 1771 when Lane was surveying this part of the Labrador coast Captain George Cartwright was further south for the second of what was to be a total of 16 summers spent in this region. Cartwright was an English gentleman, albeit a tough one, with an interest in hunting, shooting and fishing. His niece described him as 'a man of naturally strong, though uncultivated talents, of great observation, and unimpeached integrity' (Cartwright 1826). His sojourn in Labrador was motivated, in his own words 'for the purpose of carrying on various branches of business upon the coast of Labrador; and particularly, of endeavouring to cultivate a friendly intercourse with the Esquimaux Indians, who have always been accounted the most savage race of people upon the whole continent of America' (Cartwright 1792).

Cartwright did achieve one of his aims, that of building friendly relationships with the natives and trading for furs and animal oil. From a business point of view he had a lot of bad luck, and ended up losing most of what he collected. We are fortunate that Cartwright kept a meticulously detailed diary during his stay since this provides an excellent account of his activities, if frustratingly little about seabirds (Cartwright 1792; Townsend 1911). It does however, allow us to piece together some events relating to the naming of the Gannet Islands. On 27 September 1771 Cartwright was based at Lodge Bay, near Cape St Charles. His diary for that day states: 'In the evening a vessel appeared, working into our harbour I went on board, and found her to be his Majesty's brig *Grenville*, commanded by Mr. Michael Lane, who had been employed all the summer in surveying part of the coast northward of this place. I remained all night on board'.

At the time of this meeting Cartwright had not ventured much further north, and indeed his Caribou Castle (at the present day settlement of Cartwright) was not established until 1775, so it is unlikely that Cartwright knew much about the area in which the Gannet Islands lay. It seems probable, therefore, that Michael Lane or one of his crew named the islands. Gannets probably never bred in Labrador since it was much too cold for them that far north (Nelson 1978). It is possible that Lane was simply an incompetent

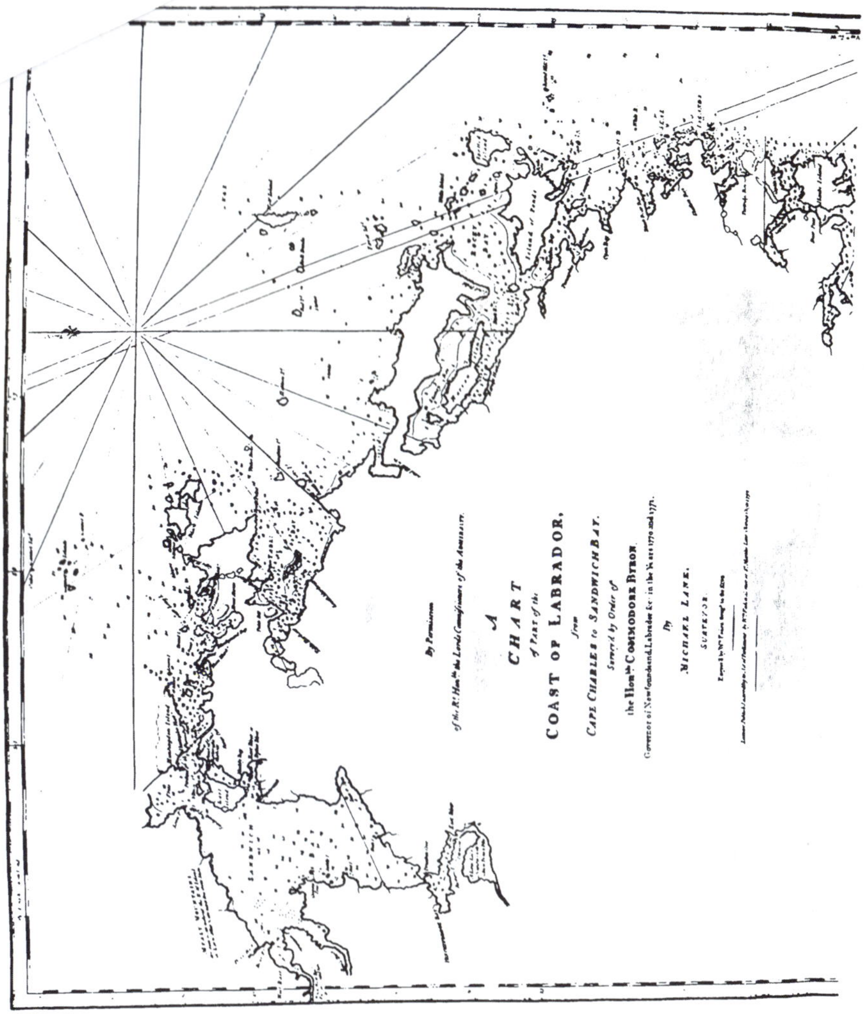

Part of Michael Lane's 1771 chart of the Labrador coast. The Gannet Islands and Outer Gannet are marked and lie at the top of the chart.

A Leach's Petrel from a burrow on GC2.

ornithologist and thought that some other seabirds (such as Common Guillemots) were Gannets and named the islands after them. However, the *Grenville*'s log makes no reference to any birds on the Gannet Islands. This is surprising since the survey coincided with the breeding season and Lane's team should have encountered auks and other seabirds had they been there. Cartwright mentions the Gannet Islands (but not seabirds) in his journal, but not until 1778, 3 years after he had moved into the Sandwich Bay area.

There are several other possible explanations for the islands' name. The one I favour most is that it stems from an observation of Gannets that occurred when Lane was charting the islands. This is plausible since we saw Gannets, usually immature birds, on a few occasions during our summers at the Gannet Islands. These birds probably originated from the nearest colony at Funk Island, which was known to be extant at that time (Cartier 1534).

Another possibility is that the islands were named after a ship. Many places on the Labrador and Newfoundland coast were named after fishing boats (Story *et al.* 1990). Interestingly, an HMS *Gannet* did visit the Gannet Islands, under the command of Lt. W. A. Chimmo, but this was in 1867, almost a century after the first charts had been made. Chimmo, like Lane, was surveying the coast, and he makes it abundantly clear that he was not impressed by Lane's ability as a surveyor. 'The whole coast is 10 or 11 miles too far to the eastward, and I can never feel that we were much indebted to Mr Lane's chart. It was found in many points simply imagination' (Chimmo 1868).

The origin of the name remains a mystery. Unfortunately the issue is further confounded by the publication in 1980 of a guide to Labrador, '*Alluring Labrador*' (Anon. 1980) which erroneously states that 'The Gannet Islands, the outer islands of Sandwich Bay, are the nesting home for thousands of Northern Gannets. . .'

* * *

Although there are no Gannets breeding in Labrador, a number of other seabird species have recently started to breed in this area. During our initial visit we discovered the first breeding Leach's Petrels and Fulmars in Labrador. It is possible that there have always been a few Leach's petrels on the Gannet Islands, but because they are nocturnal they would easily have been overlooked during previous ornithological visits. But the occurrence of the Fulmar as breeding species in Labrador is definitely a recent development and we were pleased when we discovered several pairs breeding on GC2. Among them, although probably not breeding, was a single dark phase, or Blue Fulmar. Like the Common Guillemot, Fulmars occurs in different colour forms, with the darker ones becoming increasingly common in the more northerly parts of their range. This Blue Fulmar was a long way south of its normal range.

Later we caught some Fulmars to ring them. Their big, dark eyes make them look deceptively gentle, but they can be vicious birds to deal with in the hand and their hooked beak can inflict nasty cuts. Fulmars invariably disgorge their last meal on being caught, probably as a defensive measure, or to detract their predator. One bird which we caught was certainly successful in detracting us. In order to assess what they had been eating we tried to collect their regurgitate by placing a plastic bag over the birds' head as soon as we got it. This particular individual threw up the usually oily, fishy mixture, but in addition, the vomit included a large, hair-covered piece of skin. We did a double take as it appeared in the bag, and when we looked at it more closely, it became apparent that it was a piece of cat. Fulmars are regular scavengers, and we could only assume that the cat had come off one of the coastal ferries.

The other seabird that has recently invaded Labrador is the Kittiwake. The first breeding record was of 16 occupied nests on Outer Gannet in 1972. During our studies in 1978 we found three pairs on GC4, and 48 on Outer

Fulmar chick: one of the first Fulmars to be reared in Labrador in recent history.

Gannet (see below). Over the next few years the Kittiwake population continued to expand, and by 1985 there were about 100 breeding pairs in total. The Kittiwake's prior absence from Labrador is surprising since they breed both to the south in Newfoundland and further north in the Arctic. One explanation for their sudden appearance is that these pioneers represent the overflow from southern populations which are known to be expanding. The other possibility is that marine conditions in Labrador have changed since the 1970s, somehow making the area suitable for the Kittiwake and the Fulmar (Birkhead and Nettleship 1988).

* * *

Some 5 km to the north of the Gannet Clusters, and clearly visible from the top of GC2, lies the Outer Gannet. This island, orientated in an east–west direction is about 1 km long and 200 m wide and is bounded by steep, 30 m cliffs on its south side. It had a reputation for being difficult to land on. Part of our remit was to assess the numbers of seabirds there, and it was on a bright sunny morning that we set off to do this. We were in high spirits, partly because of the excitement of setting off for the unknown, and partly because of the weather.

About 3 km out from the Gannet Clusters we ran into dense fog and Outer Gannet disappeared from view. We briefly considered the merits of going back or pushing on. With naive optimism we took the latter option, and having previously taken a compass bearing duly arrived at the island. We were impressed: as the cliffs on the southwest corner loomed out of the fog we could see that they were thickly covered with very accessible Brünnich's Guillemots and Kittiwakes. From this position however, no Common Guillemots at all were visible. A heavy swell, together with rocks that had been ground smooth by generations of sea-ice, confirmed the difficulty of landing. Moreover, we had to do more than simply get ashore, we had also to either moor the Zodiac or get it out of the water. Mooring it was out of the question; the anchor we had was minimal, we knew nothing about the currents and there were a few big lumps of ice bobbing about. With some nimble acrobatics we managed to get ashore at the base of the Brünnich's Guillemot cliffs and with rather more difficulty, and a lot of effort, we got the Zodiac into a safe location. We tied it securely down before setting off to explore. Using a tripod I took a photograph of us all just after we had arrived, joking that this would be all that people would find of us.

Our efforts to get a feel for the island were severely limited by the fog. We stood at the highest point and for a few moments as the fog cleared we were able to see the central part of the island below us, containing thousands of Common Guillemots, which up until now had been obscured. Crouching down in the grass to reduce our conspicuousness we exchanged exclamations of awe: I had certainly never seen so many Common Guillemots together (it was only later that I visited Funk Island: Chapter 4). One reason why this colony was so impressive was because the birds were breeding on essentially flat

Our first landing on Outer Gannet (l–r: Remey O'Dense, TRB and Richard Elliot).

ground, only a few metres above sea level, in a wide rocky area sculpted into irregular shapes by the elements.

Obtaining an accurate estimate of the numbers of Common Guillemots was impossible with the fog periodically obscuring most of the island, but it seemed to me that there were many more than the 17,700 pairs recorded during the previous visit, by Tony Lock, in 1972. We agreed that Richard and Remy would come back to Outer Gannet later in the season to make some counts during their tour of Labrador's seabird colonies.

As we made our way round to the north side of the island in the dense fog we were excited to find huge pieces of ice piled up on the rocky shore. At this stage I had never seen ice like this before, and I suppose its presence there fufilled some of my glamorous expectations about the rigours of Labrador. We climbed onto the ice to enjoy the feel of it, oblivious at that stage that its presence spelt trouble.

There was no ice on the south side of the island when we returned to the Zodiac and, using the compass, set a course through the fog for the Gannet Clusters. We had been going for little more than 15 minutes when we started to encounter a large quantity of ice. Initially we made light of it since it seemed only to add to the sense of adventure. We used the paddles to push the ice out of the boat's path, slowly maintaining our course towards home. However, after a while the ice became totally impenetrable and we had no option but to turn back and try another course. By this stage the potential seriousness of our situation was apparent and our high spirits evaporated, all conversation stopped and communication was limited to instructions.

The ice had depressed the swell and the sea made no sound, so the silence was immense. The fog was as thick as ever and the dreadful monotony was

Common Guillemots at Outer Gannet Island.

Brünnich's Guillemots. The bird in the foreground is incubating and its egg is just visible.

broken only by the occasional Common Guillemot flying out of and back into the gloom. In the next couple of hours we tried several other channels in the maze of ice, but with the same disheartening result. Just as things started to seem really serious and almost as though by magic, the dim outline of land appeared and we were back at the Gannet Clusters.

Once back at our camp we brewed a pot of tea and ate an enormous belated lunch, and as we did so the fog disappeared. On climbing to the top of GC2 and looking back towards Outer Gannet, the huge amount of sea-ice that had blown in from the north was apparent. It was also clear that had we spent much longer in finding our way we might never have made it back to the islands since the ice was now packed tight against the Gannet Islands and beyond right to the mainland.

Apart from a small area of open water between GC2 and GC3 there was virtually no water visible between us and the mainland. As we reflected on our good fortune, the quiet of the afternoon was shattered by the loud expiration of a whale in the nearby patch of open water. Suddenly a second whale blew, and as they surfaced we could see that they were Minke whales. They appeared to be trapped in the ice-free water within the Clusters.

As we watched, the smooth black arches of the whales' backs continued to appear and reappear as they returned to this single area of open water to breath. Then to our surprise, the whales started to stand upright in the water, with a full third of their 8 m-long bodies projecting from the surface, hanging there for a few seconds before slipping back below the surface. Their pectoral flippers

Outer Gannet Island from GC2 after the fog had lifted, showing the extensive slob ice we had to negotiate on our journey back from Outer Gannet.

with their diagnostic white band were clearly visible. The two whales continued to appear and disappear for about twenty minutes, presumably looking for a way out of the ice. I later discovered that this 'spy hopping' behaviour is not unusual among Minke whales, and that they were indeed searching for an escape route. Whether they found it or not I do not know: after ten minutes they simply disappeared.

Until fairly recently the relatively small Minke had not been considered worth hunting. However, with the near extermination of the larger whales, the few remaining nations that see fit to kill whales are concentrating on this species.

* * *

The earliest ornithological reference to guillemots or any seabird on the Gannet Islands is for 1895 when Mr Dicks 'collected a large series of the eggs of this bird at the Gannet' in that year (Macoun and Macoun 1909). Townsend (1911, p 379) suggests that prior to this Common Guillemots were abundant on the coast during Cartwright's time (1770s), but points out that Cartwright does not mention this species by name. However, Cartwright confusingly makes a number of references to shooting 'strangers', which he then defines as 'a waterfowl of the duck kind'. Townsend suggests that these might have been Common Guillemots. The terms 'strang' and 'strany' were Scottish names for the Common Guillemot or Razorbill (Lockwood 1984; see also Appendix 1), so Townsend's supposition is at least probable. On the other hand Cartwright also refers to 'murrs' elsewhere in his journal, so he apparently knew what a murre (i.e. Common Guillemot) was.

Oliver Austin (1932) visited the Outer Gannet in 1928 but found no guillemots breeding there. We do not know whether this was because there had never previously been guillemots there, or because as elsewhere (Nettleship 1985) it reflected the years of ruthless exploitation by fishermen. By 1944 both species of guillemot were certainly breeding at Outer Gannet (Tuck 1961), but it was not until Leslie Tuck's visits in the early 1950s that any reliable estimates of numbers were made. Since then the number of the two guillemot species breeding at Outer Gannet and the Gannet Clusters has increased dramatically (Fig. 13).

* * *

As I made counts of Common Guillemots on the different islands I was struck by the new opportunities that the Clusters presented in terms of understanding their biology. At this stage, in 1978, I had studied Common Guillemots only at Skomer Island and a few other locations in Britain, all of which were remnants of much larger colonies. These tiny aggregations had left me myopic and in Labrador I realized that very large, thriving colonies have a completely different feel to them. The birds seem more confident and they bred in places, and at densities, they would not do so elsewhere. This alone encouraged me to return to the Gannets, but there was a further element that in the end made it worthwhile. This was the high proportion of bridled Common Guillemots at the Gannet Islands. An important aspect of any scientific research is to exploit any unexpected opportunities that arise. And this is exactly what I attempted to do during that first, brief visit and subsequently.

The Common Guillemot occurs in two different plumage forms: normal and 'bridled'. The latter has a beautiful white eye ring and a line of white feathers running back from the eye in a narrow groove of feathers. These birds

Ice between the islands: the patch of open water between GC2 and GC3 where the two Minke whales were temporarily trapped. Our camp is at bottom right.

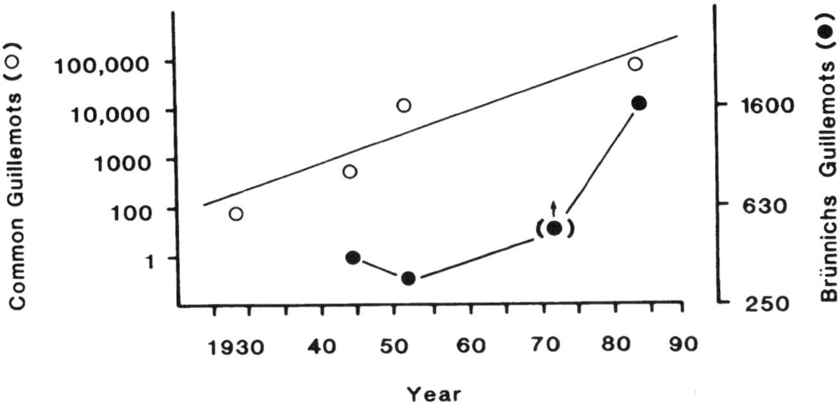

Fig. 13 *Changes in the numbers of Common and Brünnich's Guillemots at the Gannet Islands (Gannet Clusters and Outer Gannet combined) over time. Common Guillemot numbers have increased at a more or less constant rate of 5.3% per annum. Data for the numbers of Brünnich's Guillemots are less reliable, probably because at the Gannet Clusters they breed in among Common Guillemots and may have been overlooked in some surveys. The 1972 count is thought to be an underestimate – hence the brackets and arrow. Brünnich's Guillemot numbers apparently remained similar at Outer Gannet at about 400 pairs from 1944 to 1983, while numbers at the Gannet Clusters have increased during the past 40–50 years (see Chapter 9). Information from Tuck (1953), Brown* et al. *(1975) and Birkhead and Nettleship (1987a).*

were once thought to be a separate species *Uria lachrymans*—the tearful guillemot—but as soon it was apparent that they freely inter-bred with the normal birds it was clear that the bridled birds were simply a colour form of the normal Common Guillemot. The occurrence of two (or more) such forms in a population is referred to as a polymorphism. Other birds also exhibit this phenomenon, for example the blue and white phases of the Snow Goose, the light and dark phases of the Gyrfalcon, Fulmar and Arctic Skua. Among invertebrates, the common garden snail *Cepaea* is also polymorphic. The interesting question for evolutionary biologists is what allows these different colour forms to persist.

Although there were bridled Common Guillemots on Skomer, there had been so few that I had more or less ignored them. Bridled birds are absent or rare in the warmer parts of the Guillemot's range and increase in their relative abundance as one moves further north. In Portugal where the last few pairs of a once large population linger on, none of the Common Guillemots is bridled. On Skomer between 1% and 2% of all Common Guillemots are bridled, but in Shetland the figure is 28%, and at the Common Guillemot's northern limit, on Bear Island, half the birds are bridled. The distribution of bridled Common Guillemots was well known because the late H.N. (Mick) Southern had organized a number of surveys, involving large numbers of amateur ornithologists, over a 30-year period to map their distribution (Southern 1962; Birkhead 1984).

Guillemots (from top to bottom): Brünnich's Guillemot, a bridled Common Guillemot and a normal Common Guillemot on Outer Gannet.

We do not know whether the Common Guillemot's 'bridle' provides the birds with any advantage. Since bridled birds are more abundant further north, or more precisely, in regions with cooler air or sea temperatures, they may be at some advantage in cooler regions. It seems unlikely that the white bridle markings themselves confer any advantage to their bearers, but it could be that these simply act as markers for some other difference. For example, it could be that bridled birds are more tolerant of low temperatures.

Les Tuck had contributed to Mick Southern's survey in the 1950s by recording the proportion of bridled Common Guillemots at both the Gannet Islands and at Nunarsuk Island (in Inuit: 'lovely little island'), one of the most northerly breeding locations for the Common Guillemot in Canada. He recorded 32% bridled birds at the Gannet Islands and no less than 71% at Nunaksuk.

I started counting. Of several hundred birds I counted only 17% were bridled. I counted again, and I counted on all parts of the Clusters, and on Outer Gannet: the figures still came out at about 17% for the Clusters and 21% for Outer Gannet. This represented a substantial reduction since Tuck's count of 32%, and I began to wonder how such a change could have occurred.

If the proportion of bridled Common Guillemots had decreased at the Gannet Islands, I wondered what might have occurred at Nunarsuk Island further north. I was subsequently able to find out because Tony Lock visited this island, and several others in northern Labrador at the same time that we were on the Gannets. He discovered that instead of there being 71% bridled birds as Tuck had found, only 27% of the Common Guillemots at this relatively small colony were bridled. In addition, Tony counted bridled birds at several newly discovered colonies in that region (The Pyramids, The Barbican, The Castle and Kidlit Island) and found that they ranged from 11% to 35% (Birkhead and Lock 1980).

These results from the Gannet Islands and Nunarsuk therefore suggested that a big decrease in the proportion of bridled birds had occurred since Tuck's visit. However, comparing the two sets of counts was not straightforward. Simply eye-balling the percentages (71% to 27% at Nunarksuk and 32% to 17% at the Gannet Islands) certainly indicates a substantial drop. However, our assessment of how meaningful these changes are depends very much on the number of birds upon which these percentages were based. If, for example, Tuck had based his estimates on a total of just 10 birds for example, no-one would consider the estimates very reliable. On the other hand, if his estimates were based on several hundred or a thousand birds they would be much more convincing.

The problem with Tuck's published information was that there was no mention of how many birds he had checked in order come up with his values of 71% and 32% for the two colonies. I wrote to him and he replied by saying he had based his estimate on about 1000 birds at the Gannet Islands and all the birds he had seen at (*c.* 300) Nunaksuk.

This suggested that a real decline had occurred, but there was another snag. The changes at these two colonies had occurred over 25 years—a comparatively short time period, since Common Guillemots may live this long. For the decline to have occurred through natural selection, the mortality rate of bridled birds would have to have been unrealistically high. Given that the Labrador populations were also increasing at this time, this scenario seemed unlikely. Another possibility is that over the 25-year period there was differential emigration of the bridled form or immigration by the normal form. However, there is no evidence either for or against this idea and unfortunately the true explanation remains unknown.

While I was in the process of making my counts of bridled Common Guillemots at the Gannet Clusters I noticed something unusual. In those places where the rocks had eroded to form long, narrow gullies, the breeding groups of Common Guillemots formed long lines. As I counted along the line I got the strong feeling that the bridled birds were occurring in groups. For example, a line of 30 birds might contain four bridled individuals three of which were next to each other. I started to look specifically for more gullies and lines of Common Guillemots to try and check this. There were plenty of birds and I had soon accumulated information on the location of all the bridled birds in 32 rows. Later, back in Sheffield I consulted a statistician colleague and we were able to confirm that bridled birds were much more aggregated than one would expect by chance. Why? One possibility was that bridled birds preferentially paired together. I had checked the composition of 109 pairs of Common Guillemots to see if this was true. But pairing is entirely random with respect to bridling, meaning that pairs consisting of two bridled birds, or two normal birds, or one of each, occurred exactly in the proportions you would have expected if the two forms paired up without any regard for each other's appearance.

The most likely explanation for this pattern is that young Common Guillemots tend to return to breed close to where they were reared. It was not until several years later that based on observations of hundreds Common Guillemots ringed as chicks on Skomer Island, I was able to confirm that some young birds do indeed tend to return to breed close to where they were reared (see Chapter 6). Obviously, both normal and bridled Common Guillemots show the same tendency to return to their natal site, but the effect is only detectable through the bridled birds. Subsequently I discovered that in some of the very first studies of Common Guillemots the tendency for bridled birds to aggregate had been noticed by others (Jourdain 1922; Huxley 1939).

The relatively high proportion of bridled birds at the Gannet Clusters was later to play a central role in my study of the Common Guillemot's social behaviour and was also responsible, indirectly, for a change in the direction of my research interests (see Chapter 8).

* * *

The day arrived when the *Pauline-Gladys* was due to return and carry us back to the mainland. There was still quite a lot of ice on the sea so we were not over-optimistic that we would be leaving. Our counts and observations continued, but we were ready to leave at short notice should we hear the distant chug of the boat's engine. The next day the sea conditions were the same, but again we were not disappointed since the weather was beautiful and our blue phase Fulmar returned. On the following day there still no sign of Ivan Roberts. By this stage our food supplies were starting to run low, but I still avoided eating the hard tack. It rained all morning and a stiff breeze blowing off the ice meant that the air temperatures were low, so we lay in the tent in our sleeping bags and read. By the afternoon it was dry and for want of something better to do we set off to explore GC5.

During our brief visit to the second largest of the six Gannet Clusters we were able to confirm one of the principles of island biogeography, albeit in a qualitative manner. Biogeographers have shown for several groups of organisms, such as birds, reptiles and amphibia that the number of species increases with increasing size of island. There are a number of possible reasons for this pattern, but one of the most likely is simply that larger islands have a greater number of different habitats and are therefore able to support more species. This appeared to be true for the plants on GC5. We found a willowherb, Labrador tea, two species of cinquefoil, dandelions, sphagnum and reindeer moss, none of which we had encountered on GC1, 2, 3 or 4. We also found surprisingly large numbers of Puffins and about 250 Common Guillemots. During the course of the day we decided that if there was no sign of Ivan Roberts by the following morning we would set off for Grady in the Zodiac. We had tried to avoid doing this for several reasons. First, we had almost too much equipment to fit into the Zodiac, and second, it looked like a long way in a small boat with a very small (10 hp) engine.

In the event, although there was no sign of Roberts, the next morning was cold and foggy with a lot of ice and we were forced to delay our departure. I resorted to eating the hard tack: it was not all that bad as long as you had a plentiful supply of saliva to help it down. The fog hung around until mid-afternoon, after which it cleared and we finally set off. The Zodiac was loaded to the inflatable equivalent of the gunnels, and our rather inadequate outboard pushed us through the water at a gentle speed. Initially it was fine, but once away from the shelter of the islands the sea got choppier and we were pushing into a head wind. Within minutes each wave was breaking over the bow of the boat making us wetter and wetter, and colder and colder as we baled the water out of the boat. At times it felt as though the engine was hardly making any headway against the wind and sea. The only respite occurred as we entered a patch of loose sea-ice where the water was calmer. Two and a half wet hours later we reached Grady harbour and landed on Little Grady Island. I felt as though we had just set foot on another planet; I had never seen anywhere quite like it. The desolate settlement comprised a dozen or so buildings and fishing flakes on both sides of a narrow tickle. The wooden houses and flakes were devoid of any paint and some were supported on the rocky seaward side by poles. As we landed and started to unload our equipment, some children appeared from the shacks. They probably thought *we* had come from another planet. They were excited by our arrival and interested in the Zodiac. Their fathers soon appeared and they too looked at the inflatable boat in disbelief when we told them where we had come from.

Grady is the summer fishing settlement for a few 'liveyers' (see next

The settlement of Grady on Grady Island, with the Gannet Islands visible on the horizon.

paragraph) from Cartwright and fortunately we had a contact here. Wes Bird was the son of the family Richard had stayed with in Cartwright. We were shown their house across the 'tickle', and squelched our way up to it. Wes Bird's elderly father greeted us at the door, and told us we could come in to dry off, but said that because there was little room we would have to sleep in one of the abandoned buildings. Mrs Bird stoked up the boiler in order for us to dry our clothes, but in the process of getting more wood she slipped and injured her ankle. Her husband Willis, a real old timer, seemed unperturbed, and Mrs Bird simply sat with her leg up, and said in her broad, Labrador accent 'I do hope it ain't broke'. My conversation with the Birds was limited: they could not understand my English and I found it almost impossible to fight my way through their strong accents and unusual dialect. Without the assistance of Richard and Remy as interpreters, I could barely make myself understood. Labrador English sounded like a mixture of strong Norfolk and Canadian to me: grammatically erratic with some interesting irregularities.

They talked about being 'down north' meaning north, and 'up the coast' meaning south. A 'liveyer' (pronounced live ere) is a permanent resident of the Labrador coast, to distinguish them from the numerous 'floaters', who come each summer on fishing boats from Newfoundland. The narrow strait dividing the settlement of Grady in two was a 'tickle'. When it came to discussing the bird life, I had to learn a whole new vocabulary. The Kittiwake was a 'tickle-ass', Razorbills were 'tinkers', guillemots of either species were 'turrs', Fulmars were 'noddies', Black Guillemots were 'pigeons', Little Auks (winter visitors) were 'bull birds' and Puffins were 'parrots'. Of course, these names had not arisen *de nouveau* in Labrador and most of them had their origin in rural Britain (see Appendix 1).

A common porpoise drowned in a salmon net, at Grady, Labrador.

We ended up spending the night in the Bird's kitchen round the boiler, having consumed large quantities of home baked bread and jam. One needs to have endured a week or so without bread, a touch of hypothermia, plus the cruel novelty of North American commercial bread, to appreciate how good Mrs Bird's bread tasted that evening.

The Birds and the other families at Grady were there for the summer fishing season. This was their livelihood and they told us that as the fishery had declined, fewer and fewer families were doing this, hence the dilapidated houses. Several boats were moored in the tickle and there was fishing equipment everywhere. On the rocky shore there was a dead Porpoise and beside it the skin of another. I asked Willis where they had come from, and then learnt about the by-catch in salmon nets: Porpoises, seals and seabirds. I assumed that having accidentally killed these small cetaceans they might then be eaten, but apparently because they had died by drowning (rather than being shot) they were not worth eating. I wasn't clear why. The blubbery skin however made excellent fire lighters: Willis demonstrated by throwing a few cubes into the boiler which flared and crackled noisily in response.

We were hoping to catch the coastal ferry, the *Petit-Fort*, back to Cartwright, but learnt over the radio that it had been delayed. Instead, the RCMP in Cartwright decided they would come and pick us up and take Mrs Bird to hospital there to check on the state of her foot. In a little over an hour we were back in the comparative luxury of Cartwright. We found some lodgings, where we enjoyed a shower, and ate what must rate as the meagrest of meals I have ever experienced in any kind of lodging. As we ate, my eyes were drawn to a colourful but out-of-focus Christ on the wall. I could not make it out, until I got up from the table and walked across the room. The eyes of Jesus followed me. Not like a Rembrandt, but with an uncanny flicker. It was then that I realized that this was one of those double colour images usually reserved for cartoon characters on children's toys. This was a face of religion I had not previously experienced. When I mentioned it to the others, Richard told me he had actually met the salesman that peddled these contradictory religious images as he had travelled north on the coastal ferry.

* * *

Much of the satisfaction of being a field biologist stems from the detailed knowledge and understanding of one or a few species. As may well have already become apparent, the Common Guillemot is one of the species whose biology I have found most interesting and tantalizing. It therefore seemed appropriate to provide a general account of its behaviour and life style, before discussing some of our findings in Labrador in more detail. The next chapter thus presents an outline of the Common Guillemot's biology.

CHAPTER 6

Skouts, Skuttocks and Strangers

The Common Guillemot . . . although every place affording a foothold is crowded to excess, the utmost order and decorum everywhere prevail; each seems desirous of assisting and accommodating the other.

Jones (*c.* 1867)

A colony of Common Guillemots (hereafter simply Guillemots in this chapter) has a pulse all of its own, albeit an erratic one. The life-force that surges through the colony gives rise to an irregular babble of noise, movements that are at once both rhythmic and arhythmic, and, at close range a distinctive, but not unpleasant odour. The pulse comes close to stopping only occasionally when a predatory gull alights nearby. Then, the Guillemots are both still and silent, waiting with nervous apprehension. The gull, a goose-sized Great Black-back, approaches the incubating birds with an aloof and careful step. As it comes closer, part of the tension in the Guillemots is released, and they start to nod their heads downwards, uttering a low, staccato moan. Their calls alert Guillemots in other parts of the colony and a ripple of unease radiates outwards from the vicinity of the gull. Just visible under the breast of one of the outermost Guillemots a crescent of turquoise egg is visible. The gull stares, and after remaining stationary for a few moments, walks forward in a confident manner, but as it does so a phalanx of Guillemot bills rapidly thrusts out, forcing the gull to hesitate. The Guillemots maintain their dense array of jabbing spears, but the gull lunges forward with open bill in a nervous attempt to grab the egg from under the incubating Guillemot. Once more it is driven back, but the Guillemots are unnerved. The bird whose egg is the focus of the gull's attention is now frightened and stands up, exposing its egg entirely and encouraging the gull to try again. Just as the gull is about to move in and grab the egg, another Guillemot breaks free from the centre of the breeding group, advances rapidly and, like David against Goliath, launches itself pecking and beating its wings against the gull and drives it from the ledge.

It does not always end like that. Often the Guillemots fail to deter the gull, who swallows his prey, egg or chick, whole. A few gulls specialize in taking the eggs of auks, and are indifferent to the Guillemots' defences. The specialists are ruthlessly efficient: they simply grab a Guillemot by its back feathers, toss it to one side and then take their egg or chick. Observing interactions between predator and prey always creates tension, but also a sense of excitement and, depending upon whose side you are on, the outcome can elicit relief or revulsion. When the victims are embryos or new born young, feelings are likely to run high. Such emotions are unavoidable, but scientific observers must attempt to think their way around, or through them in order to analyse the events that have just taken place. Why, for example, should one member of the Guillemot colony place itself at risk by attacking and chasing off the gull? After all, it was not his, or her egg that required protection.

A simple, but entirely erroneous, explanation for this behaviour would be that the Guillemot was behaving for the benefit of the colony as a whole. This is the comfortable 'explanation' we were brought up with and one that the script-writers of wildlife television programmes so often perpetuate. However, we now know that this kind of explanation is no explanation at all because evolution does not work that way, and to see why you would need to read Richard Dawkins' (1976) *Selfish Gene*.

So, if doing things for the good of the group or colony cannot account for this apparently altruistic, anti-predator behaviour, what can? The answer to this question is, I believe, the key to understanding the social behaviour of Guillemots. We do not yet know the true answer, but we can look at other aspects of the Guillemot's biology to help us decide which of several possibilities is most likely.

The first possibility is a selfish one. That is, the Guillemot who chased away the gull was not doing it for anyone else's benefit but his own. By deterring the gull and causing it to fail in its predation attempt the gull may be less likely to return to that particular ledge in the future. The aggressive Guillemot therefore protects its own egg or chick from future attack. Luckily for the other group members they also benefit, but their gain is purely incidental.

Another possible explanation for this kind of anti-predator behaviour is that it has evolved through something called 'reciprocal altruism'. The gull-chasing Guillemot frightens off the gull now because at some time in the future another Guillemot will do the same for him or her. You scratch my gull and I'll scratch yours, later. Much of human friendship is based on reciprocal altruism and is thought to work because the individuals concerned (1) can recognize each other, so they know exactly who has and who has not helped them in the past, (2) live in more or less permanent groups and (3) are relatively long-lived, so that opportunities for reciprocation exist. In fact Guillemots also fulfil all these criteria. Judging from how Guillemots respond to different individuals on their breeding ledge it is clear that they recognize their immediate neighbours from their appearance and probably by their voice as well. They are extremely faithful to their breeding sites and the majority return to exactly the same breeding site each year. Finally, adult Guillemots can expect to live for about 20 years on average.

It used to be thought that these criteria alone were sufficient to account for reciprocal altruism, but behavioural ecologists (e.g. Koenig 1987) now consider that one additional feature must be present. It is that unselfish behaviours, such as gull-chasing, must be costly, by which we mean that it must involve some kind of risk to the Guillemot. In other words, the

individuals that perform the unselfish behaviour must suffer some reduction in the likelihood of survival or reproductive success. Without this 'cost' there is nothing to repay by reciprocation. Obviously demonstrating that such costs exist is difficult and although the idea of reciprocation is an attractive one, with the exception of our own species, there is little good evidence for it. Since reciprocal altruism appears to account for so few cases of selfless behaviour in birds (Koenig 1987) or elsewhere in the animal kingdom (Taylor and McGuire 1987), I think it is probably unlikely to explain any aspect of Guillemot behaviour.

The third explanation rests on an idea which is now accepted by most biologists, that biological success is measured by how many copies of your genes you get into subsequent generations either directly through your own reproductive efforts, or indirectly, through those of your relatives or kin. If breeding groups of Guillemots comprise groups of genetically related individuals, then gull-chasing could evolve through what is called 'kin selection'. Basically, because the gull-chaser is related in some way to the potential victim, by stopping the gull from eating its relative's egg, the gull-chaser increases its own long-term genetic success. In other words, by behaving in this apparently altruistic manner the gull-chaser will end up with more copies of its genes (including those for bashing gulls) in future generations.

The degree of relatedness between parents and their biological offspring is one half, expressed numerically: 0.5. This is because offspring obtain half their genes from their father and half from their mother. Following a similar line of reasoning, the average degree of relatedness between offspring and a step-parent would be a quarter or 0.25 and between cousins one-eighth, or 0.125. Using an extreme example in an attempt to illustrate the numerical principle of kin selection, J.B.S. Haldane, the eminent geneticist, said he would give up his life for two brothers or eight cousins.

Although we do not know exactly how closely related the Guillemots within a breeding group are, it seems certain that they comprise groups of relatives since studies of ringed birds have shown that many young Guillemots return to settle close to where they were reared. However, because parentage may not be exactly what it appears to be in Guillemots (see Chapter 8) it is not sufficient to simply record whether putative offspring return to breed near their parents in future years. Instead it will be necessary to determine the degree of genetic relatedness among birds in the same breeding group through a technique such as DNA fingerprinting.

This may be possible some time in the future, but Guillemots show several other sophisticated forms of social behaviours which are consistent with the idea that group members are related. Before describing those it might be useful to have a concise outline of the Common Guillemots' biology, so that all these behaviours can be placed within the framework of the species' life cycle.

* * *

As a research student on Skomer Island in the 1970s I marked several hundred Guillemots with different combinations of colour rings so that I could recognize them individually. The simple act of placing a unique combination of rings on a bird's leg completely and utterly changes the way one perceives them. Instead of being anonymous units they are instantly transformed into distinct personalities whose fortunes can be followed over several successive years.

Guillemots walk, or rather shuffle on their heels (strictly, their tarsi) and consequently usually wear through their colour rings within four or five years. In areas where they have several metres to walk to their breeding sites the rings tend to wear through faster than where they breed on narrow ledges and do not do much walking. It was on a narrow site like this where in 1973 I ringed a female Guillemot who was brooding her chick. Thereafter she was referred to by her ring colours: green-green-white (abbreviated to G-GW), and she and her unringed partner occupied that same site each year thereafter. G-GW was a particular favourite of mine; she was extremely confiding and each year when I ringed her chick she remained on the ledge. I was disappointed when, in 1991, she finally failed to return. Since G-GW was a successfully breeding adult when I originally caught her, she must have been at least 20 years old when she finally disappeared, presumed dead.

This was not an isolated instance, many other Guillemots also returned year after year to the same breeding site, living to a ripe old age. By knowing the proportion of adults which survive from one year until the next we can calculate their likely average life span. G-GW was one of many birds that I kept an eye on over several successive years in order to estimate the survival rates of Guillemots. And between 90% and 95% of birds return from one year to the next, showing that on average Guillemots live for 10–20 years after reaching breeding age (Fig 14). Mike Harris and Sarah Wanless have recorded similar survival figures for Guillemots on the Isle of May in eastern Scotland, as have a number of other people studying Guillemots elsewhere.

In fact the Guillemot is typical of many seabird species in this respect, the majority of which are long-lived. Interestingly, this characteristic is part of a package of features that seabirds and a few other groups of birds, such as vultures, share. All long-lived birds tend to produce rather small clutches, and many seabirds produce a clutch of just one egg. Long-lived birds also take rather a long time to reach sexual maturity. Common Guillemots first start to breed when they are about five years old on average. The Fulmar and some albatrosses, which are among the longest lived of all birds, may wait until they are 10 before they start breeding!

These characteristics, referred to as life history features, are typical of seabirds: long life and a slow rate of reproduction. Compare this with the life history features of the diminutive Zebra Finch which occupies the arid parts of Australia. Like the Guillemot these birds also breed in colonies, but their life history could not be more different. My colleague Richard Zann found that of all the Zebra Finch chicks he ringed, virtually all of them had reached the end of their life 12 months later: an almost unbelievably short average life expectancy of just 6 months! Accompanying this high mortality rate, Zebra Finches start to breed at just 2 months old. Each clutch comprises four or five eggs and they breed two or three times in rapid succession. Their life history

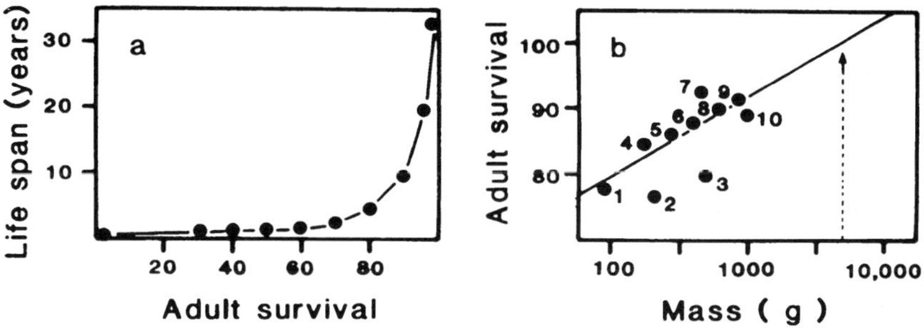

Fig. 14 *Survival and longevity. (a) The relationship between annual adult survival (the percentage of adult birds surviving from one year to the next) and the average further expectation of life. At one extreme few Zebra Finches survive from one year to the next and their average life expectancy is just 6 months. At the other extreme about 96% of Wandering Albatrosses survive between years and their life expectancy is over 30 years. (b) The relationship between body weight in auks and their annual adult survival, showing that larger species tend to have higher rates of survival on average. Somewhat ironically, this relationship predicts that the Great Auk (indicated by dotted line and arrow) would have had a very high rate of survival. 1. Least Auklet, 2. Ancient Murrelet, 3. Pigeon Guillemot, 4. Cassin's Auklet, 5. Crested Auklet, 6. Black Guillemot, 7. Atlantic Puffin, 8. Razorbill, 9. Brünnich's Guillemot, 10. Common Guillemot. From data complied by I. Jones.*

can be summarized as fast and furious: a short life expectancy coupled with an early start to breeding and a high reproductive output.

* * *

Studying Guillemots requires rather little skill, but infinite amounts of patience. Because they breed on open ledges and at traditional sites, one only has to construct a hide opposite a breeding group to be able to watch and record everything that Guillemots do. In many ways watching Guillemots is like going to the movies: you get into the hide, open the window and it all happens in front of you. The only thing the observer has to do is to know which part of the movie to record.

Guillemots are notoriously tame and will allow people (and other predators) to approach very closely before they panic and abandon their egg or young. For this reason, they used to be referred to as the 'foolish guillemots' (Morris 1856), or in German as *trottellummen*, meaning idiot bird. However, their apparent tameness is not stupidity, nor does it mean that they are unaware of a human observer sitting out in the open. Being concealed in a hide ensured that the birds continued with their normal behaviour, and it also protected me during bad weather. The only time being in the hide was unpleasant was during the occasional hot summer's day. On those occasions, a sun-roof would have been a blessing.

There are few birds whose social behaviour in all its aspects can be observed and documented so easily, and I have always enjoyed the spectacle provided by Guillemots. Not everyone however would agree that this is an ideal research situation. Field ornithologists can be divided into two broad categories, 'watchers' and 'doers'. I am definitely a watcher and get a buzz from simply seeing and recording what birds do. 'Doers' are those that cannot sit still, they prefer to be always on the go and they try to record data using methods that do not involve sitting still for long periods. So a 'doer' would not sit in a hide for hours checking to see which birds have laid; instead he or she would dash in, make a spot check on all the birds, then rush off to measure something else, before coming back to make further spot checks during the day. The information obtained using these two different approaches may well be similar, but to my mind the 'doer' misses out on what makes birds tick. This is a subtle point and I am sure doers would argue that since they obtained essentially the same information, they know what makes a particular species what it is. However, I suspect I am after something more: I want to try to get inside the mind of the bird. In a modest way, I liked to think that the thousands of hours I sat watching Common Guillemots on the Gannet Clusters in Labrador resulted in a degree of understanding of their behaviour, if only for a short time, that was similar to that achieved by people like Cynthia Moss with African elephants, or Fiona Guinness with red deer on Rum (Moss 1988; Clutton-Brock *et al.* 1982).

Simply watching and recording what animals do does not result in good research. Biologists must use the approach that they feel is most appropriate, and use it to test specific ideas. This can be achieved only by having a good theoretical background to the subject, and knowing what questions to try and answer. Only in this way do they know what is worth recording in their note book.

* * *

The bed in which you sleep is probably well designed and acceptably comfortable. The settee that you sit on to watch television is also likely to be agreeable. However, ask a designer to produce a bed-settee and they are forced to start comprising, and the end result is either okay for sitting on, but impossible to sleep on (or *vice versa*), or uncomfortable for both sitting and sleeping. The Guillemot is like a bed-settee in that it is a compromise or, biologically speaking, it is the result of conflicting selection pressures. On the one hand a Guillemot has to be able to fly in the air, but it also has to be able to swim underwater. In achieving this compromise the Guillemot has had to trade performance in one against the other.

When Guillemots travel from their breeding colony to a feeding area they are fine just so long as they can fly in a straight line. Ask a Guillemot to do anything else and it is in trouble. This can be witnessed all too easily at the breeding colony. On a calm day a Guillemot returning from a fishing trip approaches its breeding colony fast and low over the sea. At about 50 m from the cliff the Guillemot alters the angle of its wings allowing it to sweep rapidly upwards until level with its breeding site. The bird then stops flapping, and plops down directly on to its site. Perfect! On a blustery day however, the birds can appear pitifully inept, rising either too fast or too steeply and occasionally colliding with the cliff wall. And like some unlucky cartoon character, they slide down the cliff-face desperately trying to regain some aerial control before hitting the rocks at the bottom. Fortunately Guillemots are robustly built and they rarely appear to suffer from such mishaps.

The reason Guillemots appear so incompetent is their wings have evolved both to propel them through the air and to fly through the water. Because water is a much denser medium than air (try clapping your hands under water), the Guillemot requires a relatively small wing for underwater 'flight'. The area of the wing has therefore been reduced during the course of evolution, so that the Guillemot has among the smallest wings for its body size of any bird. One consequence of this is that it has to flap extraordinarily fast simply to remain airborne, which means it has a rather rapid flight (78 km/h), but a distinct lack of manoeuvrability.

In fact, the Guillemot's small wings are still rather large for underwater locomotion, and if you watch a Guillemot swimming under water, say from a cliff-top or in an aquarium, you can see that it holds its wings partly closed to reduce their size even further while diving. The thin film of trapped air which overlies its plumage gives the Guillemot a mercurial lustre as it effortlessly twists and turns underwater in pursuit of its prey.

As well as having speed and agility Guillemots have the power and physiological ability to dive to considerable depths. Dives of 1 min duration are routine and the longest recorded dive was 2 min 20 s. Auk biologists have utilized a variety of approaches to try and work out how deep Guillemots and other seabirds can dive. For example, one observer saw a Guillemot surface with a flatfish, which was presumed to be swimming on or close to the bottom 60 m below the surface. John Piatt and David Nettleship (1985) recorded Guillemots in Newfoundland drowned in fishing nets set as deep as 180 m! Subsequently, using ingeniously designed micro-depth gauges, Alan Burger (1991) recorded Guillemots diving to 138 m in the Witless Bay area of Newfoundland. The Guillemot and other members of the auk family do not appear to show such extreme specialization for diving as the penguins, yet the maximum dives of the Guillemot and other large auks are as deep as those for several penguins (Figure 15). This figure also shows something that was predicted long before there was any information on seabird diving depths, namely that larger birds are capable of much deeper dives than smaller ones. Considering the auks of the north Atlantic, the maximum dive of the Little Auk (Dovekie) (which weighs about 160 g) is just 35 m, the Atlantic Puffin about 60 m and the Razorbill and two guillemot species well over 100 m. There are several reasons for this relationship. First, bigger birds have larger oxygen reserves and so can remain submerged for longer. Second, larger animals are stronger and hence physically more capable of overcoming the problem of buoyancy that affects small diving animals.

Diving to these maximum depths must be energetically demanding, and of course the deeper the bird has to dive to find food, the less time there will be for actual foraging once they are down there. Birds will therefore attempt to

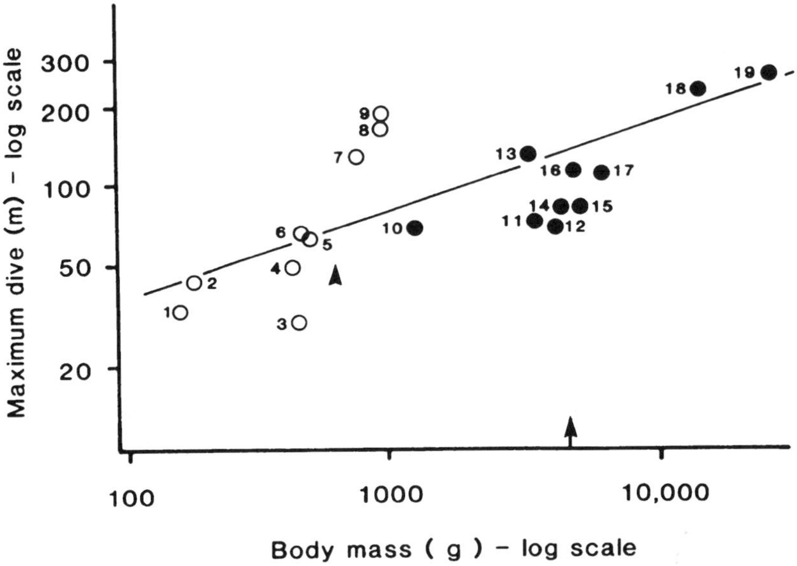

Fig. 15 *Relationship between body weight and maximum diving depth for auks and penguins showing that heavier birds are capable of deeper dives. This figure also indicates that the Great Auk (arrow) would be capable of diving to a depth of at least 100 m (see text). 1. Little Auk, 2. Cassin's Auklet, 3. Pigeon Guillemot, 4. Black Guillemot, 5. Rhinoceros Auklet, 6. Atlantic Puffin. 7. Razorbill, 8. Brünnich's Guillemot. 9. Common Guillemot. 10–19. Penguins: 10. Little, 11. Magellanic, 12. Chinstrap, 13. Jackass, 14. Humboldt, 15. Adelie, 16. Macaroni, 17. Gentoo, 18. King, 19. Emperor.*

forage in the most economical way, and this means feeding in fairly shallow water whenever they can. As an illustration of this, Alan Burger (1991) found that although Brünnich's Guillemots are capable of diving as deep as 153 m, most of their foraging was done at depths of 20 or 30 m. There is no equivalent information for Common Guillemots, but several other species show a pattern similar to that of Brünnich's Guillemots, so it is likely that Common Guillemots search for food underwater in a similar manner.

* * *

One of the most striking features of the Common Guillemot's life style is how very closely they pack themselves onto their breeding ledges. Why do they do this, and what are the consequences of breeding so close together?

The first question amounts to the same as saying: what is the adaptive significance of high density breeding? Or, why has this kind of high density living evolved? No other species breeds in such close proximity, and even the Brünnich's Guillemot does not form such dense groups (see Chapter 4). Other cliff nesting species, such as the Kittiwake, Fulmar and Razorbill, at least space themselves out so that each pair has its own separate domain. So what is it about Guillemots that makes it worthwhile for them to breed so close together? As indicated in the opening paragraphs of this chapter, high density breeding probably provides Guillemots with some protection from predators.

This is an easy idea to test, at least in principle. All that is required is to compare the reproductive success of birds breeding at different densities. My first attempt to do this, on Skomer, was relatively crude. I selected a few ledges on various parts of the island where Guillemots bred at different densities and then laboriously recorded their breeding success (see Chapter 7). The results I obtained showed that on average about 80% of those birds in the highest density groups successfully reared a chick to fledging compared with only 30% of those at the lowest density. Subsequently Tony Gaston in his study of Brünnich's Guillemots on Prince Leopold Island (see Chapter 7) developed a more sophisticated technique which involved considering, not just a few breeding ledges, but each pair's individual breeding site. The advantage of this technique over mine was that it avoided the possibility that there might be something peculiar about an entire breeding ledge which might affect the results, and hence his approach was more reliable.

I adopted Tony's technique in Labrador and found that Guillemots breeding with several close neighbours, i.e. at high density, were significantly more likely to rear a chick to fledging than those with fewer adjacent neighbours (see Chapter 7). Later, Ben Hatchwell (1991) used this same technique on Skomer and confirmed that the two approaches gave similar results.

The avoidance of egg and chick predation by gulls and other predators is one reason why birds in groups do better, but there are also other reasons. An egg that rolls away from an isolated bird must have a much greater chance of being lost than one rolling away from a Guillemot in a dense group. In the same way, chicks in dense groups are much less likely to wander into danger than those in sparse groups. Another possible but as yet untested explanation for why breeding success is higher in dense breeding groups is that the birds in closely packed groups are just better quality individuals. If this was true it could mean that the correlation between reproductive success and breeding density is an artefact, and that bird quality rather than breeding density explains the differences in breeding success. There are two reasons for thinking that there might be something in the bird quality idea. First, young Guillemots breeding for the first time generally breed on the edges of established groups and hence have fewer neighbours and generally rather poor breeding success. Second, I have always been amazed at how difficult established breeders make it for young birds to join a breeding group. Older Guillemots are always extremely aggressive towards young birds trying to establish a breeding site of their own. Because I knew that being part of a dense group increased the chance of rearing a chick, I could never understand why established birds should be so reluctant to allow other Guillemots to join their group. However, the older birds' aggression may be part of a test, which allows only the strongest recruits to become established. It would be no good allowing any young wimp to breed next to you, since you might have to rely on him or her standing its ground when confronted by a big bullying gull. If only the toughest birds get to breed in dense groups then bird quality could contribute to the higher breeding success observed there. Overall, I suspect that if we were ever able to measure it we would find that both quality *and* breeding density independently helped to determine breeding success.

* * *

Despite its obvious advantages to Guillemots, living in ultra-close proximity to one's neighbours is not easy. There's no privacy, there is always someone in the way, and there's a constant risk of your offspring getting mixed up with those of your neighbours. Because the benefits of high density are so great, natural selection will have favoured those individuals best able to cope with such crowded living conditions. Let us start by considering the lack of privacy. When a pair of Guillemots copulate it is no secret and the entire breeding group or colony can see and hear what is going on. If it were not for the fact that some males try to exploit the copulation behaviour of their neighbours to their own ends (see Chapter 8), there might be no problem with this. On the other hand, there might actually be some advantages to such openness and for the following reasons. The spatial compactness of a breeding group is only an effective defence against predators if the entire group breeds at the same time. (Anyone breeding early or late would be particularly vulnerable to predators). Indeed, most Guillemots within a breeding group do lay their eggs over just a few days, and this ensures the effectiveness of their crowding. Synchrony is always greater within groups than between groups in the same colony, so whatever determines synchrony occurs at the level of the ledge and not the colony as a whole. The question is, how do they synchronize? What cues do females use to decide when to lay their egg? The entire process of egg formation, from the moment the female starts to lay down yolk in one of her ovarian follicles, to the time the egg is on the ledge, takes about 2 weeks (Birkhead and del Nevo 1987). Notwithstanding, most females on the same ledge lay their egg within a few days of each other. One possible answer to how they manage to tune their timing of egg laying so finely, is that they actually use the sight and sound of other colony members copulating. Obviously this is no conscious process (Guillemots do not think 'Aha, there's so and so copulating, we'd better get started'). Instead, those individuals whose physiology allowed them to respond to the sight and sound of copulating neighbours in such a way that meant that they laid at the same time as most other birds, would have left more surviving offspring than those which did not respond in this way. Although no-one knows exactly what cues colonial birds use to synchronize their breeding activities, it is striking in Guillemot colonies how the sight and sound of one copulating pair can trigger off a wave of mating by neighbouring birds (Tuck 1961; Birkhead *et al.* 1985).

Although this sounds plausible, there must be more to it than this, and I continue to be fascinated by the detailed mechanics of how synchronous egg laying is achieved in colonial birds. Even if all group members decided to start yolk production on exactly the same day (and this is unlikely), differences in the quality of individual birds would inevitably disrupt the synchrony so that some birds would be ready to lay their egg while others were still finishing it off (see Fig. 23 in Chapter 8). This probably means that certain birds 'hold back' their egg, while others might accelerate the completion of theirs. If you measure the dimensions of eggs in a Guillemot colony you will find that the variation in egg size is enormous: the largest are 30% bigger in volume than the smallest. This variation in egg size may be part of the synchrony mechanism, and a bird that 'knows' (so to speak) that it will be late might complete the production of its egg sooner by producing a slightly smaller egg, in order to enjoy the benefits of being in phase with other members of the group (Birkhead and Nettleship 1982).

We are always much more likely to forgive the peccadillos of our friends and relatives than those of strangers, and forming friendships is one way that Guillemots ensure the peaceful cohesion of their breeding group. Using the term 'friendship' to describe relationships between animals sounds dreadfully anthropomorphic, but it implies only that some sympathetic bond exists between the participants. This is exactly what there appears to be between certain Guillemots. A being from another planet might deduce something about who our friends are by recording how much time we spend with different people and how we behave towards them. Using a similar technique we can do the same for Guillemots. Within their breeding group each pair of Guillemots has a fixed breeding site, so their social options are limited to their immediate neighbours. Within this tight circle of acquaintances, Guillemots have clear favourites, and their relationships are reflected through mutual preening (or allopreening, as its called). Birds breeding in colonies spend a considerable amount of their time together preening each other. Their preening activity is usually focused on those parts which they cannot reach themselves: the head and throat. Allopreening probably started off as a behaviour in which one bird removed parasites, such as ticks (see Chapter 7 for life cycle), from another. This behaviour might not have been entirely selfless since there would be advantages in ensuring that your neighbour was free of such invertebrates, but the recipient would certainly benefit from having parasites removed from those otherwise difficult places. It is therefore easy to see how allopreening could be useful in forming a bond between two individuals. It is also straightforward to see how such a behaviour could become ritualized, by which I mean allopreening could continue to serve as a bonding behaviour even if there were no parasites to remove. It may be sufficient, as it is in humans and other primates, that the mere process of being preened or groomed by another is a pleasurable sensation. I have rarely seen parasites on adult Guillemots, indicating that allopreening between pair members must be either entirely effective, or largely ritualized. However, on the few occasions I have seen birds with ticks they elicited a very strong preening response from other birds. Some indication of the importance of allopreening within pair members can be gauged by how much it occurs. Throughout the time that a bonded pair of Common Guillemots are together at the breeding site they preen each other at least once each minute and for several seconds at a time. This works out at a total of well over 100 hours of allopreening contact time each season!

Just as preening between pair members helps to develop a bond between them, birds with adjacent breeding sites also allopreen each other. By carefully recording who allopreened who I noticed that Guillemots do not allopreen their neighbours at random, instead certain individuals clearly preferred each other and preened each other more than others. In mixed colonies of Common and Brünnich's Guillemots in Labrador, where the two species often bred side by side (see Chapter 7), either species will allopreen the other. Friendly behaviours between species are somewhat unusual, but here the advantage is clear, since by forming such relationships, they help to keep the entire breeding group together, and protect it from predators.

Many highly social species use preening or grooming in this way, but what we don't know yet for Guillemots is to what extent these friendships we see involve genetic relatives. Given that young birds may return to breed close to where they were reared themselves, it is entirely possible that the close social relationships among certain Guillemots involve relatives.

* * *

Several years ago when I was en route to Australia to study Zebra Finches, I spent several days in Singapore. I am sure there are more crowded places on earth than Singapore City, but I had never seen quite so many people before in my life, and as I made my way through the congested streets, I found myself apologizing to everyone I bumped into. Most of the locals did not of course, they were used to their populous city, but I was using conventions that I had been brought up with. In crowded circumstances and especially in an unfamiliar country, we are often at pains to make our apologies as explicit as possible, presumably because we are concerned that someone might misinterpret our actions as aggressive and respond antagonistically. It worked and I survived without a single unpleasant encounter.

Being in a Guillemot colony is like being in Singapore: a Guillemot can barely turn its head without the risk of antagonizing a neighbour. The Guillemots' situation is a tricky one since they need to maintain the exclusive occupancy of their tiny breeding territory, but at the same time they do not want to weaken the fortress of their group defence. The way they get round this is by having a repertoire of different apologies, or appeasement displays. For a Guillemot that encroaches on another's territory retribution is swift and sharp. An aggressive Guillemot lunges out and delivers a penetrating blow with its pointed bill. But it then immediately performs an appeasement display as though to say: 'that's it, I've made my point, and now no more!'. The recipient may retaliate with a mock lunge, but then it too will apparently change its mind and signal its appeasement. Such is typical of Guillemot interactions; appeasement is rife and only rarely do full scale fights occur. The appeasement signals Guillemots use take a number of forms: simply turning the head away, a rapid in-out stretching the head and neck away from an aggressor, and preening the flanks. The latter display, which I called side-preening, caused some confusion among early ornithologists. Edmund Selous, the zealous and meticulous 19th century observer of birds, thought that Guillemots must be infested with fleas to break off from their fights so often in order to preen themselves (Selous 1901).

* * *

On the east coast of England just a few kilometres from the holiday resorts of Bridlington and Filey there is an impressive stretch of sheer limestone cliffs, which in places reach a height of 150 m. In the late 1800s, the farmers who owned the land adjacent to the cliffs harvested the eggs of Guillemots and other seabirds that bred there. Working in gangs of three and using ropes as thick as a man's calf, they would lower one of the gang over the cliff. With a linen bag slung over each shoulder and armed only with a stick tipped with an iron hook, the man on the end of the rope would collect as many eggs as there were within his reach. He would then be hauled back to the cliff-top and lowered down again further along. A man could collect several hundred eggs on a good day, and thousands were taken each year. Fortunately, a number of writers saw fit to document the activities of the 'eggers', or 'climmers', as they were known, so we now have a record (Nelson 1907; Wade 1907) of what went on.

The eggs were used for local consumption, others were sent to Leeds for use in the patent leather industry, or were used in the clarification of wine, and there was always a ready collector's market for unusually coloured eggs. The commoner varieties of eggs were sold for a penny or less each, but the better-marked ones could fetch as much as 7s 6d each (37.5p in present currency, but probably equivalent to £20.00 at today's prices). The range of colour, pattern and type of marking among Guillemot eggs is remarkable and some collectors specialized in this species, making large display cabinets solely of Guillemot eggs. Because exceptionally coloured eggs were valuable, the climmers tended to remember where they had found them. It did not take them long to realize that the same distinctive eggs appeared on the same ledge, at exactly the same spot each year. 'Certain peculiarly marked eggs have been taken from the same ledge for twelve or fourteen successive years. . .' As Wade (1907) realized, this indicated two things: that female Guillemots lay a similar coloured egg each year and that they tended to use to the same breeding site each year. He wrote: 'A well-marked egg is found year after year on the same ledge of rock as long as the bird lives . . .'. Considering the general level of knowledge about the biology of birds at the end of the last century, Wade's account of the Guillemot's breeding cycle is extraordinarily accurate.

The fact that female Guillemots always lay eggs of the same colour is not unexpected. Most birds probably do, but this combined with the fact that each female lays a uniquely patterned egg, puts a different complexion on the subject. The most obvious explanation for this variation is that it allows females to recognize their own egg. Given the proximity of neighbouring Guillemots within a colony the risk of eggs becoming muddled up is relatively high, and being able to identify your own egg would enable a bird to retrieve it. This would be important, in an evolutionary sense, since it would avoid individuals wasting time and energy rearing someone else's offspring. Natural

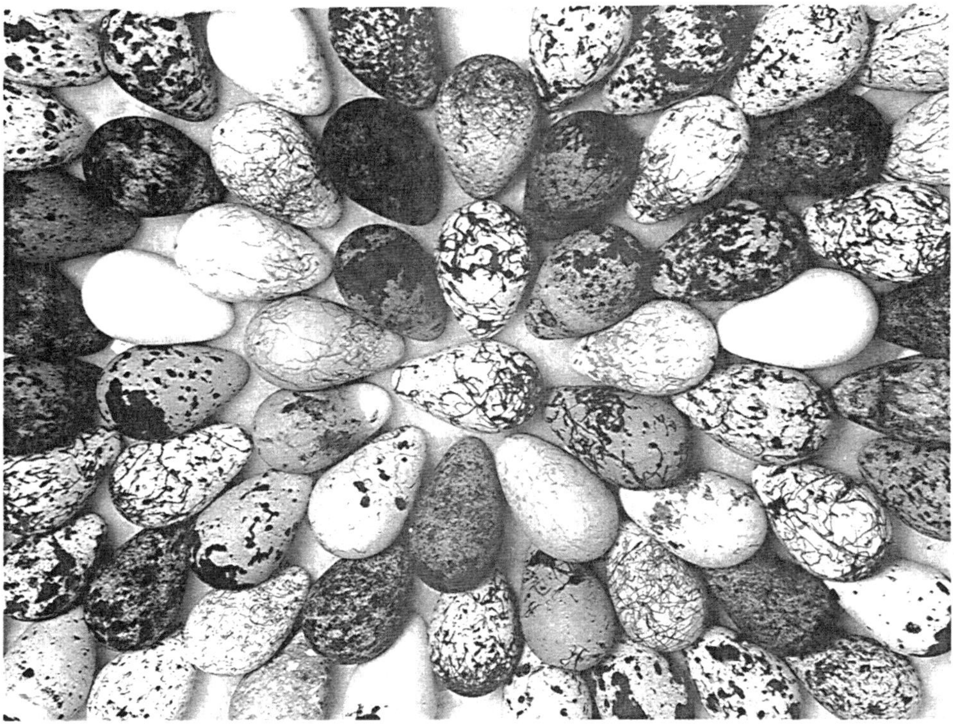

Eggs of the Common Guillemot showing the incredible variation in tone and makings (Photo: D. Hollingworth. courtesy of the University Museum, Manchester).

selection would therefore have favoured individuals that laid distinctive eggs and had the ability to recognize them. The recognition process would have been aided by the fact that females produce eggs whose colour and type of marking are consistent from season to season.

These ideas were tested by the Swiss ethologist Beat Tschanz (1959, 1968) in a detailed series of experiments on the steep cliffs of the Lofoten Islands, off the coast of Norway. He found that Common Guillemots were indeed able to recognize their own egg and they did so by using the two most constant features of the egg: its background colour and the type of marking. The actual distribution of markings on the egg surface, which can vary quite a lot between eggs laid by the same female, is not used in the recognition process. This sounds like a straightforward story; the female lays an egg with the same colour and pattern each year and she learns to recognize these features. Her partner also learns what her egg looks like, and either of them will retrieve the egg if it rolls away from the breeding site. However, the Guillemot's behaviour is slightly more sophisticated than this indicates. Breeding ledges are notoriously grubby places, largely because Guillemots are completely indifferent about where they defecate. During dry weather this does not create any problems, but after rain there can be an abundance of slippery guano on the ledges and eggs inevitably change their appearance. Tschanz's experiments showed that if the colour of the egg is changed over a few days, either by the experimenter, or naturally as result of it becoming soiled, then the parent birds can cope with this and modify their perception of what their own egg looks like.

These results indicate that the Guillemot's ability to recognize its egg is an adaptation that allows it to enjoy the benefits of breeding at high densities and still keep track of its own egg. If making sure you care for your own egg and no-one else's is important, then recognizing your own chick must be just as crucial. Tschanz also studied this aspect of the Guillemot's behaviour and made some interesting discoveries. A day or two before the start of hatching the chick breaks through the membranes inside the egg to puncture the air space at the blunt end of the egg, and from this point on starts to use its lungs for respiration. It is also at this stage that the chick starts to communicate with its parents. It is uncanny to pick up a Guillemot egg and hear the peeping of the unhatched chick inside, but these calls, and the answering cries of the parents, are essential parts of a mutual recognition system. By the time the chick has hatched several days later, the parents and chick have learnt to recognize each other's calls. From the moment the chick first pierces the shell until it emerges usually takes about 2 days, but sometimes as long as 5 days. This is partly because the Guillemot's egg shell is relatively thick, an adaptation to protect it in the absence of a soft nest. Tschanz found that the calls of different adults and chicks vary considerably, but subtly, such that all are individually distinct, allowing both parties to recognize each other even before

Five-day-old Common Guillemot chick.

Twenty-day-old Common Guillemot chick – ready to fledge.

the chick climbs out of its shell. Mutual recognition subsequently plays a vital role in the survival of the chick on the ledge, and later ensures that the chick and its father remain in contact during the dangerous process of fledging (see below).

The extent to which the Guillemot's egg and chick recognition system is an adaptation to crowded living conditions can be seen most easily by making a comparison with a closely related species that breeds at lower densities. The Razorbill is considered to be one of the Common Guillemot's closest relatives (Strauch 1985) and never breeds as close together as Guillemots, always leaving at least one body length's distance between themselves and their nearest neighbour. There is therefore no immediately obvious reason why the ability to recognize either eggs or chicks should be as well developed in Razorbills as in Guillemots. Indeed, if we were to find that Guillemots and Razorbills did not differ in this respect it would be convincing evidence that egg and chick recognition was nothing to do with living in crowded conditions.

One has only to compare the eggs of the two species to realize immediately that those of the Razorbill are much less variable in colour and markings than those of the Guillemot, suggesting that the scope for Razorbills to identify their own eggs would be limited at best. The eggs of the Razorbill generally have a white or cream ground colour and blotchy markings which range from

David Quinn

Razorbill.

reddish brown through to black. Only once, in Labrador, have I ever seen a Razorbill egg with the beautiful 'pencil' markings that occur regularly on Common Guillemot eggs. I always regret never having had the wit to photograph this apparently unique egg.

How does one check to see if Razorbills can recognize their own egg? The answer is, very simply, by giving them a choice between their own egg and that of another Razorbill. Within the discrete nest site of a given pair of Razorbill's I moved their own egg a few centimetres from where it was normally incubated, placed another Razorbill egg equidistant from the incubation spot, and allowed the birds to come back and make a choice about which one they wanted to incubate. I then returned an hour or two later and simply felt which of the two eggs was warm to record their choice. I performed this test with 40 pairs, and my expectation was that if they could not tell the difference between their own egg and that of stranger, then in about half the cases the Razorbills would choose their own egg. And that was pretty close to what I found: 20 pairs retrieved and incubated their own egg, 17 took the strange egg, and in three cases the birds incubated neither egg (Birkhead 1978).

Paul Ingold, who was one of Tschanz's students, studied Razorbills in much more detail than I did, and he thought that Razorbills could recognize their own egg (Ingold 1980). I was rather concerned by this difference between our results, but when we met and discussed this discrepancy it transpired that in his studies, Razorbills could identify their own egg only when it was very different from the one they had to choose from. Allowing for this difference between our studies, it is clear that egg recognition is less well developed in

Razorbills than in Guillemots. In the same way, the Razorbill's ability to recognize its own chick is less developed than the Guillemot's. Whereas in the latter species parent–offspring recognition occurs before hatching, it is not until the Razorbill chick is several days old that parent and offspring are able to recognize each other's calls. But why do Razorbill adults and chicks need to recognize each other at all if they do not breed on densely packed ledges? The answer is that while the Razorbill enjoys the privacy of its own breeding site, it is still part of a breeding colony, often containing tens or hundreds of other Razorbills and at fledging time it is just as crucial for Razorbill parents and offspring to recognize each other as it is for Guillemots.

* * *

If a Guillemot chick survives the first few days after hatching, then it has an excellent chance of making it through to leaving the colony at about 21 days of age. I was concerned therefore when one afternoon in Labrador I noticed that the 18-day-old chick from site 143, that I had seen earlier in the day, was missing from its site. One of its parents was there, sitting in a typical brooding position, but when it stood up I could see that there was no chick present. Since the chick had disappeared during the day I had to assume the worst, but I was disappointed and surprised because big chicks did not just disappear and were rarely taken by predators.

As I continued to scan the study plot, checking off which other chicks were present, an adult bird came in from the sea carrying a fish, and walked up to site 143. The partner at the site greeted it noisily, and the bird with the fish leant forward expecting a hungry chick to emerge from under the other parent's wing. Of course this did not happen, but the parents' behaviour was not unusual. Birds which have lost their chick often continue to bring fish back to the site for several days. But what happened next certainly was unusual because from under another Guillemot about half a meter from site 143, a chick appeared, walked over to site 143, took the fish and then snuggled up under one of the adults there. I could not believe my eyes. I had never seen chicks being brooded by birds other than their parents like this. On checking through my note book I found that the other site at which chick 143 had been cared for was one where the parents had lost their own chick about 1 week earlier.

It is not exactly true that I had *never* seen chicks brooded by birds other than their parents because after an attack by a predator, such as a gull or raven, chicks left alone by their parents would huddle under any remaining adult birds. Also, if a chick fell from its site onto another ledge below, one or more adults would brood the chick, often keeping it safe until its own parent could return it to their own site. However, what I had not seen before was an undisturbed Guillemot chick being cared for by another bird.

When something unexpected occurs like this, one can react in several possible ways. You could just shrug and assume that there was a temporary mix up, and think no more about it. Or, you can experience a slight visceral twinge of inadequacy, as I did, and wonder whether you might have been

missing something important. I quickly recovered from my uncertainty and told myself that I had, after all, spent up to 6 hours every day for the past 6 or 7 weeks checking the 255 pairs on my study plot, and I had seen no other incidents that made me suspect I had missed anything. However, in the days that followed I looked very carefully for any apparent anomalies.

That night I lay in bed unable to sleep, thinking about what I had seen. My mind oscillated between trying to decide what information I needed to collect, and what the implications of that information were. If the sharing of parental duties turned out to be a regular occurrence then I was optimistic that it formed another small piece of the Guillemot jigsaw puzzle. As my thoughts ran freely but at random in the half-way house between wakefulness and sleep, a Skomer memory emerged. I recalled an instance where I had seen a chick hatch and then grow up at a site only to disappear at about 12 days of age. However, the next day when I checked the site the chick was back under its parent's wing. I remembered being surprised by the disappearance and subsequent reappearance of this chick, but dismissed it as a mistake in my observations. I now realized that this had probably been a case similar to that which I had witnessed here in Labrador earlier that day. I subsequently discovered that almost 80 years earlier Edmund Selous (1905) had also and puzzled over a case of alloparental care of Guillemots on Shetland.

Over the next few weeks I saw several further cases of chicks being brooded by other Guillemots. Superficially, the behaviour of these other Guillemots resembled the 'helping' behaviour which occurs in several co-operatively breeding species, such as bee-eaters, some Australian wrens, babblers and the Florida Scrub Jay (e.g. Stacey and Koenig 1990), but with one important difference. Co-operative breeders operate as a group, with a breeding pair assisted by one or more non-breeding helpers. There was nothing in the Guillemots' behaviour to indicate any kind of obvious bond between the true parents and the assistant brooders. Since I had to refer to the behaviour as something, and wanted to avoid possible confusion, I refrained from using the term 'helping' and decided to refer to it as alloparental behaviour. This is a term that had been used to describe similar 'aunting' behaviour in mammals (Riedman 1982).

Careful observation that season showed that a total of 16 different chicks, out of a total of 193 that subsequently fledged successfully, received some alloparental care (about 8% of all chicks). I also noticed that all instances of alloparental care were initiated by the chick rather than by the other adult. The chick simply emerged from under its true parent, walked up to half a metre towards the other bird, and pushed itself under the alloparent's wing. The alloparent always appeared happy to accept the chick and brooded it as though it was its own. The chicks that sought and received this extra care were usually close to fledging, being 19 days old on average, but ranged from 12 to 28 days.

Most of the alloparents (54%) were birds whose own breeding attempt had ended in failure. Others were adults whose own chick had fledged successfully (23%), non-breeding birds (14%), or even in a few cases, birds which had a chick of their own (9%), which meant that they sometimes brooded two chicks at once—one under each wing! Although most alloparents were failed breeders, only about 20% of Guillemots that lost their own egg or chick became alloparents. Each chick usually had only one set of alloparents, indicating that they had a specific relationship with one particular pair of Guillemots other than their parents. The chicks used their alloparents on and off over several different days, with the longest relationship I saw lasting 10 days.

The care that alloparents gave was restricted almost entirely to brooding, and on only two occasions did I ever see alloparents feed chicks. However, in one case of alloparental care observed by Sarah Wanless and Mike Harris (1985) on the Isle of May, Scotland, the alloparents fed the chick regularly. I could not help feeling that with more co-ordination the alloparenting system in Labrador could have been so much more efficient. In the year I first saw the alloparental behaviour, some Guillemots were having great difficulty in finding food for their chick and were occasionally forced to leave them unattended at the colony when they (reluctantly) went off to forage. It is extremely unusual for Guillemots to leave their chick alone at the colony, indicating that food must have been very difficult to find that year. If the behaviour had been better organized both parents could have foraged while they left their offspring with the alloparent, but it did not happen like that.

You might be asking: what is so special about one Guillemot brooding

another's chick? One answer is that very few other seabirds would behave in this gentle manner. Most species either ignore each other's offspring or they are positively aggressive towards them. Gannets, for example are notorious for lashing out with their bayonet-like bills towards any misplaced offspring (Nelson 1978), and the larger gulls such as Herring Gulls and Great Black-backs regularly kill and eat their neighbour's chicks. Interestingly, however, gulls do sometimes adopt strange young, but the way it occurs is rather macabre. If a Herring Gull carries another's chick back to it's own territory as a potential meal for its own offspring and the meal is still alive and kicking when it drops it in front of its own young, the adult gull becomes confused because it is quite incapable of distinguishing the strange chick from its own. Not wanting to risk eating its own young, the gull ends up adopting the stranger (Parsons 1971; Graves and Whiten 1980). The only seabirds other than Guillemots where any kind of alloparental behaviour occurs are the Emperor and King Penguins. In the latter species, unattended chicks are brooded by failed breeders, just as in the Guillemot (Stonehouse 1960).

A simple explanation for why Guillemot chicks are brooded by birds other than their parents would be that it is a case of mistaken identity by both parties. But this seems highly improbable, given the fact that adults recognize their own egg and their chick (see above, and also later), and that chicks are usually being brooded by their parent when they go off to seek their alloparent. To account for this unselfish behaviour it is necessary to look at the situation from the point of view of both the chick and the alloparent and for each one consider the costs and benefits. What do chicks gain from being brooded by someone other than their parent? The answer is probably very little. At the stage when chicks seek alloparents they are quite capable of maintaining their own body temperature, so are unlikely to take or get much body heat from their alloparent. The chick does gain protection from potential predators, but no more than it could from its own parent. An intriguing, but completely hypothetical possibility is that by forming a relationship with another pair, chicks are looking to the future. Knowing birds other than their parents might just increase their chances of becoming established in that part of the colony at some future date.

What about the alloparent? Again, the costs and the benefits of alloparenting both appear to be fairly minimal. There is no obvious disadvantage in brooding someone else's chick. The benefits of doing so, on the other hand, may be may be difficult to detect. One possible advantage is that it is all part of the system that helps to bond the members of breeding groups together.

* * *

The occurrence of several unselfish behaviours, such as chasing predators, allopreening neighbours and caring for each other's chicks, all suggest that the Guillemot's social life is highly sophisticated and may rest on the fact that breeding groups comprise aggregations of related individuals. As I have said, there is some circumstantial evidence for this: bridled birds tend to occur closer together than we might expect by chance, rare egg types often occur side by side on the same ledge, and young birds often seem to settle close to where they were reared. What we need to do now is to make DNA fingerprints of all group members to see just how closely related individuals are, and try to work out the costs and benefits of these unselfish behaviours to see if they could evolve through kin selection.

* * *

Early naturalists knew that Guillemot chicks fledged before they were full grown, but had no idea how the tiny, flightless chick got safely from its breeding site, often on enormously high cliffs, onto the sea. For a long time it was believed that the chick was carried to the sea either on its parent's back, in its parent's bill, or between its legs (Cordeaux 1872). Part of this mystery lay in the fact that fledging almost invariably occurs at dusk, in the kind of light where your eyes play serious tricks with you, and it's difficult to make out what's happening. However, Edward Hodgson, a 'climmer' from Bempton Cliffs, in Yorkshire, knew exactly how they fledged, as Wade (1907) relates: 'He has seen young Guillemots and Razor-Bills called off the ledges by their parents, always at high tide, when, with rigid wings, and feet spread out on each side of the tail, they slant away to the water'.

'Fledging' is actually an inappropriate term to describe the Guillemot chick's departure from the colony since it implies that the young bird can fly. Young Guillemots are incapable of sustained flight when they leave the breeding colony, but they can, however, glide rather well on their primary coverts. The primaries, or main flight feathers of the wing, do not start to grow until some time after the chick has left the colony.

The departure process can be protracted: on some evenings the chicks look as though they are sure to go, as they leap up and down on the spot, whirring their tiny wings. Later the tiny chick walks cautiously to the edge of the ledge and, bracing itself, crouches low as if about to leap out from the ledge. But each time it 'chickens out' and waddles back to the security of its site and its parent. Such hesitation and indecision is amusing to watch, and looks mildly incompetent, but in fact is probably the opposite. This is a momentous leap, a one-way ticket, there is no going back, so everything must be just right. During those evenings of apparent hesitancy, the chick is probably making decisions that significantly alter its chances of survival. Only when everything is perfect—the state of the sea, the wind, the number of other Guillemots leaving, and so on—is it worth leaping. I suspect it is *this* that explains why some chicks leave at 18 days of age and others hang on until they are 27 or 28 days old. In fact very few chicks wait until they are more than 24 days old, but those that do so look curiously out of place, over-sized and with a plumage that

Among the best known of the climmers from Bempton, Yorkshire, Edward and John Hodgson.

no longer looks immature. Moreover, they are too big to be brooded by their parent, and usually sit or lie alongside them at the site.

Although both parents may be at the colony with the chick when it does finally decide to leave, it is the male, and not the female, that accompanies the chick out to sea. Why this should be so is not clear. The same pattern occurs in the Razorbill and in Brünnich's Guillemot.

In all studies of Guillemot 'fledging' it has been found that chicks avoid leaving on stormy nights. Obviously, heavy seas and bad weather would make things difficult for the young Guillemot, increasing the risk of being swept away, and decreasing the chances of the chick and its father meeting up on the sea.

In one year in Labrador, stormy seas delayed fledging for several days. Knowing when most chicks hatched and that chicks usually left the colony at 21 days of age, I had predicted on which night most chicks would leave. The big day came, but a huge Atlantic swell, which had built up throughout the day, meant that by the evening the sea was totally unsuitable for 'fledging', and indeed no chicks left. Sea conditions were no better the next day, or the next, and consequently chicks of 'fledging' age started to build up on the ledges. We predicted that when it did occur fledging would be highly synchronous and this turned out to be true, but in a manner that was both spectacular but totally unprecedented in my Guillemot experience. The morning of the third day was bright and clear, but with a sea that continued to pound the rocks at the edge of the colony. It was not until mid-morning that we noticed that the sea was finally starting to calm down.

Around lunchtime someone watching the Brünnich's Guillemots told me that a chick had fledged. I did not think much about it: chicks occasionally mistime their departure or are accidentally knocked from the ledge by an incoming bird. However, when they said that several others looked as though they were about to go, I thought that this was something peculiar to the Brünnich's Guillemots. I was anxious to record the fledging weights of these chicks and decided to try and exploit this unusual opportunity.

I carefully climbed down to below the Brünnich's Guillemots' breeding ledges, jumping over slippery rocks and dodging waves (something I would not have contemplated doing at dusk or dark, when the guillemot chicks usually fledge). It was a strange sensation being so close to the breeding ledges. Also, I felt so oversized. I had stared at these cliffs from my hide for hours and hours and in my mind I had decided how big they were. Now I was here, I realized that one's mind plays tricks and that the cliffs were actually much smaller than I thought.

The position I climbed to, armed with a hand-net and note book, was perfect. The beauty of this particular Brünnich's Guillemot breeding area was that the fledging chicks landed on a rocky shelf before running down to the sea, where their father was waiting for them. Here I could momentarily way-lay them, snatching up the chicks in the net in order to weigh and ring them quickly before releasing them. The whole process took only a couple of minutes and they were then reunited with their father. I was delighted, not only by the opportunity to obtain some information that I had thought would

be impossible to get, but also because I was virtually part of the fledging process. I felt the awful thump as the chick landed on the rock shelf some 15 m below its breeding site, I could also feel the eyes of the anxious father, who was sitting on the sea, burning into me as I put the chick into the small net bag to weigh it. And it was with a mutual sense of relief that the ringed chick clinked its way across the rocks into the sea to be reunited with its father.

During the hour or so that it took me to catch and deal with about 20 chicks, I was aware that the sea was becoming calmer and that the volume of noise from the surrounding Common Guillemots was increasing as they too started to fledge. This was utterly remarkable. In 10 years of watching Common Guillemots I had never seen young birds leaving in the middle of the day. Fledging normally occurs in that half light where you cannot be sure that your eyes are telling the truth, a frustrating experience when one is desperate to see some of the details of interactions between the adults and chicks, but here it was all happening in bright sunshine, and what is more, the number of Common Guillemots fledging was overwhelming. I estimated that about 16,000 fledged on that one afternoon, just over half of all the Common Guillemot chicks at the Gannet Clusters.

The spectacle was remarkable, not least because both the atmosphere and the distribution of the birds at the colony was like nothing we had seen before. The noise was incredible, with adult birds uttering their harsh, heart-rending fledging call, while the young birds replied with their high pitched and piercing '*weeloo, weeloo, weeloo*'. The calls given by males to encourage their chick is like no other, and reflects the desperate sense of urgency that at least I feel, as I watch the young Guillemots fledge. Fathers and offspring made their way to the water's edge, climbing over boulders, leaping from ledges, and finally accumulating in small groups before braving the surf. On the sea the birds swam away from the colony just as they would have done at dusk, but for the whole of that afternoon the area between the islands was full of birds. There were rather few attacks by gulls, suggesting that it is the synchronous nature of the Guillemot's fledging, rather than departing at dusk, as they normally do, which makes it safe.

* * *

The fact that most of the Common Guillemots at the Gannet Clusters bred on flat rocks and often close to sea level meant that unlike the Brünnich's Guillemots, and indeed unlike most Common Guillemot colonies outside north America, few of their chicks had to launch themselves off a vertigo-inducing cliff to fledge. This had several major advantages in terms of acquiring information about their breeding biology. In particular it meant that it was relatively easy to catch chicks as they left the colony and record their

The young of Common Guillemot (top), Brünnich's Guillemot (middle) and Razorbill (bottom), just prior to fledging. The young Common Guillemot goes to sea in what is effectively adult winter plumage whereas the other two chicks will leave in summer plumage.

body weight. The weight of a chick at fledging can provide a useful index of how good the food supply during its development has been. If food is scarce chicks are fed relatively little and they fledge at low body weights. On the other hand if food is abundant, chicks are well fed and heavy at the time of fledging. It seems reasonable to assume that chicks that are very light at the time of fledging have a poorer chance of surviving to reach adulthood than well-fed, heavy chicks. This is, however a moot point among Guillemot biologists (Gaston *et al.* 1983). Weighing chicks just as they leave the colony has several advantages over other more traditional study methods. First, it avoids the disturbance and bias that regular weighing of chicks creates. Several studies have shown that visiting Guillemot chicks every other day in order to weigh them can alter their biology: such chicks leave the colony a day or two earlier, and a few grams lighter than 'control' chicks that have not been disturbed (Birkhead 1976; Harris and Wanless 1984; Hatchwell 1990). We also found this effect for Razorbills at the Gannnet Clusters. Moreover, in order to obtain a 'fledging' weight one usually has to visit study chicks over several days until they finally disappear, and then use the last obtained weight as the fledging weight.

By catching chicks as they fledged and recording their weight, these biases were avoided. GC1 had the perfect location for doing this: the adults and their chicks had traditional fledging routes, taking them the easiest way through the rocks down to the sea. On GC1 there was such a route where I could conceal myself among the rocks, so as not to disturb the fledging process, and as the birds came round the corner I could sweep the chick up in a small hand-net. Just as with the Brünnich's Guillemots, the accompanying male bird usually rushed away and sat on the sea as I ringed, weighed and measured the chick.

Because there was insufficient room for two people to remain concealed I usually worked alone when weighing fledging chicks at this particular site. If the Guillemots saw me they simply would not come past the ambush point. What usually happened was that I would be dropped off on GC1 in the early evening and then someone would come back a couple of hours later to pick me up. There was invariably an initial delay until the first birds started to appear, but I always found this a tranquil time. I enjoyed the opportunity to sit and look at the islands from an unfamiliar position, to watch the Razorbills kleptoparasitizing the puffins far out at sea, flying up behind them from underneath and snatching the dangling capelin from their bill.

Suddenly I would be snapped into action as the first Guillemot chick waddled past my hiding place. As soon as the parent bird saw me it fled, skittering across the rocks and into the sea. The chicks were usually slower off the mark and I was able to catch them in the net. But if I missed on the first attempt, they were not worth pursuing since once they started running across the 5 m slab that led to the sea I could only rarely catch up with them: I was perpetually surprised at how fast these tiny birds could run!

This system worked well—except for one particular night. Whenever any of us visited another island, we always went prepared to be abandoned, for whatever reason, and took with us some extra food, usually a sleeping bag and waterproofs. The disadvantage of this of course was the difficulty the extra equipment made to leaping out of the Zodiac onto slippery rocks in a heavy swell, and climbing up the rocks to a secure foothold. This particular evening I was dropped off at the landing spot at the far end of GC1, and hid the rucksack full of emergency equipment in the usual place under the cliff, well above the high water mark. This was a perfect Labrador summer evening: relatively calm seas, a few ragged clouds contributing to a complex sunset and the mainland lying as a thin grey band on the horizon. I arrived at the catching place hot and sweaty, and started to organize my ringing equipment and note book when suddenly an early Guillemot chick appeared round the corner almost beside me. I grabbed the net and lunged at it, but the chick ran nimbly past me. Instead of running out across the slab, as usually occurred, the chick ran behind me and into a large rock pool of dirty green water. As the chick scrambled out of the other side I made a leap across the pool to try and turn it back. Halfway through the jump I realized that I was not going to make it right across, and resigned myself to getting my feet wet. This would be worth it if I caught the chick, but my calculated gamble failed miserably. As my feet went into the water, instead of stopping about ankle depth as I had imagined, I continued to sink and finally came to rest in chest-deep, filthy green soup. The Guillemot chick got away, and my next thought was not to spend too long in this ridiculous situation in case I missed the main rush of birds. As I climbed out onto dry land the rich odour from the pool and myself became apparent: the pool was almost entirely run-off from the adjacent Guillemot colonies and consisted of a super-concentrated solution of Guillemot guano and green algae. I undressed and wrung out my clothes. It was not unpleasant since the sun was still shining and it was a warm evening. No sooner had I reached my natal state when the first Guillemot chicks started to appear. Hastily I pulled on my shirt and started work. Fortunately this was a long work-shirt which protected my 'lap' as I dealt with the Guillemot chicks I had caught.

I had processed only a handful of chicks when I noticed with disbelief a few spots of rain on the note book. Looking up at the sky I saw heavy black clouds, plus the sickening yellow light that heralded a storm. Standing up and looking towards the mainland I could see it coming, a column of swirling grey, accompanied by white-horses on the sea. My present state was not one in which I wished to endure a storm, and I soon realized that my situation was potentially dangerous. Reluctantly I abandoned any idea of further catching, and by the time I had packed my ringing bag and put on my wet clothes it was virtually dark, raining hard and most remarkably, the sea was being whipped up into a frenzy as the squall hit the Clusters. I hurried down to where I had hidden the rucksack and got into the survival suit, effectively an insulated boiler suit. Although this would (apparently) ensure my overnight survival, it rendered me virtually immobile. I picked out the torch (flashlight) and clambered to a spot where I could see the cabin on GC2, and started signalling. No response. Through the blurred image of my wet binoculars I could just make out the camp, but no sign of anyone about and certainly no-one heading down towards the boat. I looked at my watch: despite it being completely dark it was 40 minutes prior to our arranged pick-up time. I had visions of the rest

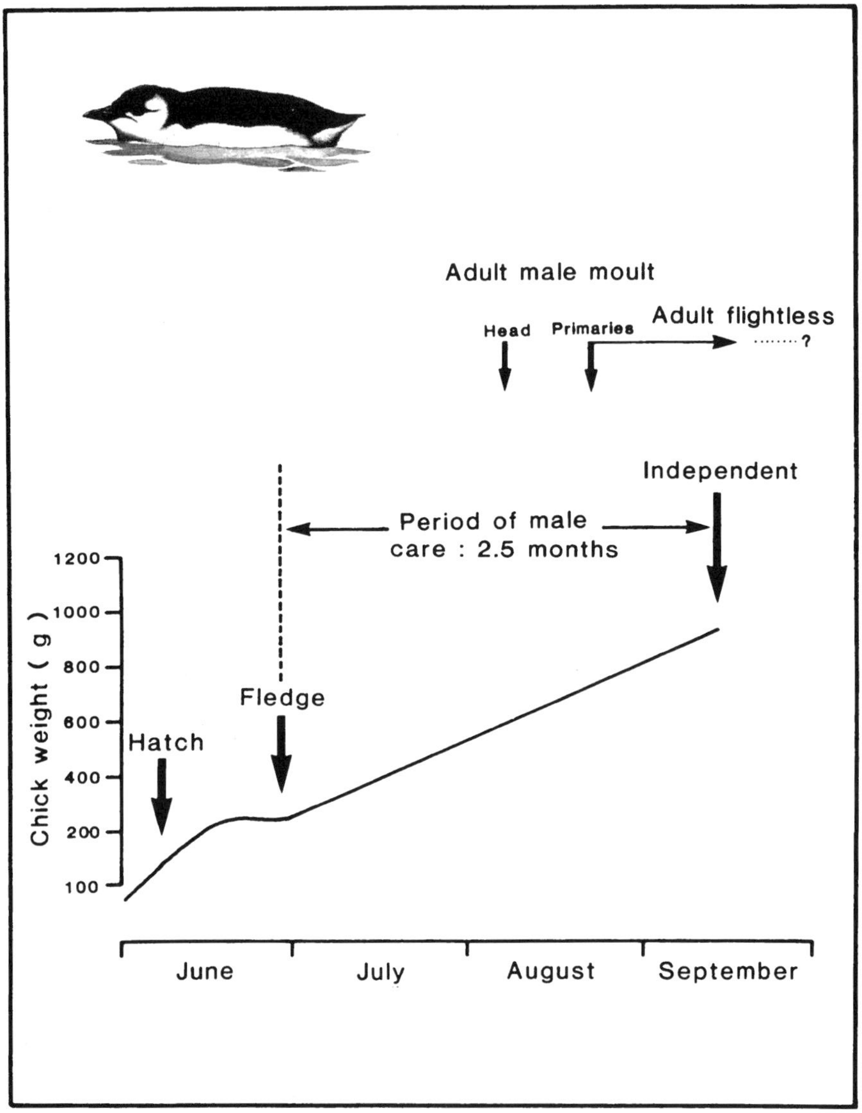

Fig. 16 *The 'fledging' strategy of the Common Guillemot, Brünnich's Guillemot and the Razorbill is unusual. The chick leaves the colony at just one quarter of adult weight and still only partly grown. Because these species breed on open sites one member of the pair has to remain with the chick constantly, to keep it warm (at least initially), but more importantly to protect it from predators. As a result, unlike the semi-precocial auks, such as the Puffin, only one member of the pair can forage for the chick at any one time.*
This figure shows the situation for Common Guillemot chicks at a British colony. Chicks typically grow rapidly, and increase in weight by about 15 g each day. They start out at 70 g at

Above: Early spring view from the camp on Coburg Island. Landfast ice in the bay: the dark line is open water, and the 2 figures are walking back to camp after caching the Zodiac.

Below: From the seabird colony on Coburg Island looking south.

Above: Razorbills and heavy seas on the Gannet Clusters.

Above: The team at Coburg *(left to right)* Keith Clarkson, Bill Carson, Don Reid (with Bruce) and Tim Birkhead.

Opposite: An Atlantic Puffin on the Gannet Clusters.

Cartwright, the largest settlement on the Labrador coast.

Funk Island: once the breeding grounds of Great Auks, now occupied by Gannets and Common Guillemots.

Ice in the bay beside the camp on Coburg Island.

Opposite above: Gannet Clusters: looking across from GC2 to GC3 at sunset.

Opposite below: A Common Guillemot, bridled form, on the Gannet Clusters.

Above: Coburg's magnificent cliffs: the pale areas are Brunnich's Guillemots, the orange areas are lichen.

The seabird cliffs near Cape Hay showing that environmental conditions can change over just a few days *(Photo: E. Greene)*.

of the team sitting warm and dry inside the cabin, playing cards, oblivious of the conditions outside.

I sat and waited. The rain drummed down on my hood and dripped down my face. The pick-up time came and went. After waiting and hoping, the reality of my situation became apparent. Despite the survival suit, which would prevent hypothermia, I began to realize that this could turn out to be a very long and uncomfortable night. I started to search round for some sort of shelter, but the rocks were unbelievably slippery and it soon became all too apparent that survival suits were designed only to be used on non-slip, horizontal surfaces. There was no shelter, so I made myself as comfortable as possible and resigned myself to a long night. Not long afterwards however, I saw the welcome flashing of torches on GC2: I responded enthusiastically with my own light and waited for the boat. I got out of the survival suit so I could climb to a suitable point and jump into the boat. After 15 minutes the boat appeared, lit by a torch, pushing its way through the rain and now heavy sea. As the boat approached the shore we exchanged shouts: clearly it was impossible to complete a conventional landing. By shouting we were only just able to make ourselves understood against the sound of the sea and wind. When Richard in the boat indicated that it was too dangerous to come close inshore in case the boat was damaged I had visions of them going back without me and having to get back into the survival suit. I climbed to a point where the rocks fell vertically into the sea, and signalled that Richard should bring the boat into this position. Once the boat was there I leapt, and landed heavily, but thankfully in the boat. Almost before my feet had touched the Zodiac's floorboards Richard had opened up the throttle to drive the boat skilfully away from the dangerous rocks. He then took me to the far, sheltered side of GC4, and dropped me off while he and the others returned to the cabin on GC2. Within an hour I had exchanged my wet and very smelly clothes for a dry set,

---

hatching, and increase to about 200 g by 10 days old. However, once they reach this stage, one parent cannot deliver food fast enough to sustain this rate of growth, and from 10 days old the chick's weight increases more slowly. However, the chick continues to mature and its wings continue to grow, and it duly fledges at about 21 days of age, at about 260 g.

Once on the sea the chick continues to be cared for by one parent (the male), but now the parent can take the chick to feeding areas, and as result is able to feed it at a high rate again, and the chick's weight increases by about 15 g per day – the same as that during the earliest phase of its life after hatching. By the time the chick becomes independent at c.100 days it is close to adult size.

The Common Guillemot's fledging strategy has evolved as a result of a trade-off between survival and growth. Compared with being on the sea, the colony is a relatively safe place to be for a chick, but obviously a chick cannot remain at the colony for ever. On the other hand, as far as growth and development are concerned, after the age of about 10 days (above) the sea is a better place to be than the colony. In terms of its chances of surviving to breeding age Ydenberg (1989) calculated whether a chick would be better off at the colony or on the sea at a given weight and age. His model showed that taking both survival and growth in to account, up to the age of 21 days the chick is better off at the colony, but after that it is better off on the sea. Thus the fledging age predicted by Ydenberg's model, of 21 days, is exactly the same as that observed in nature. Data from Birkhead and Taylor (1977); Harris et al. *(1991).*

the squall had passed and the sea returned to its earlier calm beneath the last stages of a fine sunset.

* * *

Normally, guillemots and Razorbills fledge at dusk, making use of the fading light to synchronize their departure. For predators the fledging period can be one of rich pickings, but the auks attempt to minimize their risks by fledging over as short a time as possible at twilight and also over as few days as possible. Many were the August evenings in Labrador when I would return to the cabin after watching the last Guillemot chicks fledge, to lie in bed with the door wide open. An ethereal green aurora would be starting to develop in the sky and I would listen to the duetting chorus of father and chick Guillemot pairs receding into the night as they swam south on the Labrador current.

We have relatively little detailed information about what happens to Guillemot chicks after they leave the colony, but I think we now know enough to piece together the likely sequence of events. In the morning following fledging one never sees adult–offspring pairs on the sea anywhere near the colony. Only travelling by boat can one encounter them, and their move away from the colony is probably to avoid further predation by gulls. As soon as the young Guillemot lands on the sea it is a capable diver, but its endurance is limited. A gull will force father and chick to dive repeatedly until the chick is exhausted and becomes a vulnerable meal. After just a few days at sea, however, the chick's stamina and proficiency improve enormously, and it is capable of catching plankton and slow moving fish. However, for up to 12 weeks after fledging the chick's father feeds it and remains as chaperone.

Some time after leaving the colony the adult Guillemots start their moult, and like ducks and a few other groups of birds, they shed all their flight feathers over a short period of time. They are then flightless for a few weeks, and, one would imagine, both inefficient and desperately vulnerable to predators. But it seems that the adults start to moult only once they are in an area where fishing is likely to be good. As far as their feeding is concerned, the lack of primary feathers may even be an advantage under water, and at the very least is unlikely to reduce their foraging efficiency (see Chapter 4). About 2 weeks before the flight feathers are shed the adults also begin to replace the rest of their plumage and their appearance starts to change. Their brown head plumage is replaced by white feathers, so that in winter they resemble the chicks they are still looking after (see Fig. 16).

Young Guillemots grow rapidly once at sea, in part at least because they can be fed much more frequently than they were at the colony. At the colony much of the parents' foraging time was taken up by flying, often several tens of kilometres, to and from the feeding grounds. Moreover, because the chick must always be guarded by a parent, only one adult at a time can forage for the chick. The limitation of the parents' ability to feed their chick at the breeding colony may be why the two guillemot species and the Razorbill take their chicks to sea when they are only one-quarter grown and still flightless (Birkhead 1977; Ydenberg 1989). After fledging however, the male can take his offspring to a rich feeding area, enabling it to develop rapidly (Ydenberg 1989; Varoujean *et al.* 1979; Harris *et al.* 1991). When the chick is about 100 days old, the male parent abandons it, allowing it to make its own way in the world. By this time the young bird will have grown its flight feathers and the older birds will probably have started to regrow theirs. However, despite this freedom, the recoveries of Guillemots ringed as chicks tell us that this is the most dangerous part of their life, and the time when a large number of young birds die. It is not known for certain whether mortality is greatest before or just after the young birds become independent of their fathers and are trying to fend for themselves.

By the time the young Guillemots are looking after themselves they have travelled down the Labrador current as far as eastern Newfoundland, and it is in this area, or slightly further south, that the young birds and adults spend the winter. These seas are rich in capelin, and the young Guillemots, as well as the adult birds, grow fat during the winter months. If they can avoid the Newfie hunters in their powerboats they have little to worry about, but the dangers are very real. As many as 900,000 guillemots of both species (but mainly Brünnich's) are shot each winter off the Newfoundland coast (Elliot 1991).

In the spring when the adult birds start to make their way north again to the breeding colonies in Labrador the young birds remain in Newfoundland waters. Guillemots in their first year of life have never been recorded visiting breeding colonies. Even in their second summer, the young birds may visit their natal colony only for a few days late in the season. No-one knows how they find their way back, but we can be fairly sure that during their first 3 weeks of life on the breeding site, the Guillemot chicks fix all the topographic features that surround them firmly in their mind so that later they can recognize when they've reached home. The behaviour of 2-year-olds is always hesitant: they approach the colony on the water, swimming nervously round the rocks where the older non-breeding birds are congregating at the edge of the colony. They might not even get out of the water, and may remain at the colony only for a day or two, before returning to the open sea. The following

year, as a 3-year-old, they are slightly braver and join the other young birds in the 'clubs' on the tidal rocks beside the colony. They may even attempt to form a pair bond, although they will probably do this many, many times before finding a suitable partner with whom to breed. Not until they are 4 years old will the skouts, skuttocks or strangers, as Guillemots were once known, have the confidence to start looking at potential breeding ledges, and it will not be until they are at least 5 years old that they get round to breeding for the first time. Only then will the cycle of life be complete.

CHAPTER 7

Between species in Labrador

The ecologist watching the populations may well not see them competing severely although the biogeographer has strong evidence that competition must sometimes occur.

MacArthur (1972)

After 3 years elsewhere in the Arctic the opportunity arose to fulfil at least some of the potential that the Gannet Clusters in Labrador offered. In the interim the project had expanded further than I might have wished, and there were seven of us that met in Cartwright one cold, wet June day in 1981. The project, funded by the Canadian Wildlife Service, was to compare the breeding and feeding biology of all five of the Gannet Clusters' auk species. I felt slightly uncomfortable about the magnitude of this project, not least because training six novices could result in my spending a lot of time making sure they were collecting good data while I did not. As it turned out I need not have worried since most of the people involved in the project over next few years were exceptionally competent and highly motivated.

It was decided that Eric Verspoor and I should go on ahead and the others would be delivered by a Sikorsky S61 helicopter the following day with all the equipment and supplies. For this first season we were fortunate to have Sid Elson, from Cartwright, with us. Sid was about 50, and his role was to assist with setting up the camp and to help us commute between the islands that made up the Clusters.

In grey and depressing weather conditions we set off for Grady, with Eric and I in the Zodiac and Sid following in his clinker-built boat. This first part of the trip was relatively easy because the numerous islands and headlands protected us from the weather and the sea. We took a break at Grady, where Sid showed us the rusting remains of the whaling station in which his father had once worked. He also introduced us to the pleasures of pink partridge berries that had fermented over winter on the bush beneath the snow. However, once we struck out from Grady to the Gannet Islands, which we could occasionally see through the murky rain clouds, my sense of well-being evaporated. The seas were mountainous, and at the top of a swell one could see the Gannet Clusters, but at the bottom, only sea. I had (almost) complete faith in the Zodiac, but I recall looking back at Sid and his boat several hundred metres behind us and thinking what an insignificant speck they made on a vast, unfriendly sea.

In due course we arrived at the Gannets. We made a makeshift camp on GC2 where the base camp was to be, choosing a site which was level, free of

The Gannet Clusters in 1981. Our camp on GC2 is visible in the foreground, with GC4 in the middle distance, GC5 behind that and the mainland of Labrador barely visible in the far distance.

birds and which provided clear views of most of the other islands. The next day the helicopter dropped the others and the equipment off on GC6. Although dumping the supplies on GC6 was less convenient than taking them directly to GC2, we did this so that the helicopter would not disturb the birds which were already incubating their eggs. Ferrying our huge pile of supplies from GC6 to GC2 was Sid's job, and no easy one in his small boat. Much later he confided to me that when he saw what was expected of him he thought we were all mad. Fortunately, he also said that since he had never been defeated by any task, he simply got on with it.

By the time the mission was eventually completed we had erected our Parcol, for cooking and eating, and several tents for sleeping. Our inexperience in selecting suitable accommodation rapidly became apparent. The Parcol was designed for the cold and snow of high Arctic conditions, but not for the prolonged rain we had in that first summer. The Parcol leaked. The tents had been chosen on the basis of their suitability for garden parties, and had several huge windows covered only with fly netting. They too leaked when the rain blew horizontally (which it did rather often). It got worse: the food had been selected by someone who obviously considered school dinners to be the ultimate in *haute cuisine*. Niggled, but not despairing, we got on with the large number of other jobs that needed to be completed before we could start fieldwork. Once our accommodation was sorted out the next priority was to erect our hides around the edge of the cove on GC4. These at least were well designed and, being made from sturdy plywood, were actually more comfortable than our base camp, despite their small size.

Main camp on GC2 in early 1981 showing the Parcol (that leaked) and the tents (that let the rain through their huge windows).

The team in 1981 (l–r): Sid Elson, Ted Currie, Randy Milton, TRB, Ruth McLagan and Eric Verspoor (top). The team in 1982 (l–r): Eric Verspoor, Richard Elliot (and Raven), Sue Johnson, Andrew Mcfarlane, Keith Clarkson (and Raven), and TRB.

When we had first arrived at the Gannets there were three large icebergs among the islands and one, estimated to be about 60 m tall, became grounded just off GC2 near the camp. We soon got used to its towering presence, but one morning just after we had landed on GC4 the berg started to disintegrate in a noisy, slow-motion manner. With a series of spectacular explosions the pointed iceberg cascaded down upon itself until there was nothing left except myriads of small pieces washing around on the sea.

We soon stepped into a routine: up at 07.00 h for a cooked breakfast, after which Sid ferried some of us the 500 m across to GC4, so we could start our observations. We remained there until lunchtime after which we returned to GC2 to work on Razorbills and Puffins. The two guillemot species, or murres as I had to refer to them with my Canadian colleagues, were my priority. In many respects they are the most difficult of the auks and it was important to start training field assistants straight away in the niceties of our established study methods. 'Murre' is an old local name for the (Common) Guillemot, which became the standard name for this species in the New World (see Appendix 1). Les Tuck's (1961) book *The Murres* had been my bible while I was conducting my PhD. However, at that stage the name 'murre' was alien to me, and like all Brits I had trouble pronouncing it. I tried various alternatives: 'murray' (to rhyme with 'hurry'), 'mure' (to rhyme with pure), but the correct pronunciation, which is onomatopoeic, rhymes with 'fur', albeit with a deep Canadian growl.

Training the numerous assistants I soon discovered that only certain personality types were suited to sitting for long periods, waiting for guillemots to stand up to provide a glimpse of their egg. Some assistants were absolute naturals and after just a few hours of instruction were totally reliable. Others were less so. One particular assistant raised doubts in my mind even before he got anywhere near a hide. As we set up our camp I had helped him carry his personal luggage up from the boat. He had an enormous, rather macho, holdall. I could barely carry it and had to drag it up to the site of our camp. When he joined me I jokingly asked him what was in it. He proudly (and naively) opened it for me to see: the bag contained nothing but paperback books, dozens of them. I stared in disbelief; he explained that he 'knew' there would be a lot of time with nothing to do on such a remote place.

I was not really surprised then, when one morning as I went from hide to hide checking that everyone was getting on OK with their new task, this particular assistant did not open his door when I knocked. I banged again on the door, but still there was no reply. On pulling the door open I saw him sitting at right angles to the window, eyes closed, rocking back and forth in time with the music which was blasting out of his headphones, high as a kite on dope and oblivious of my presence.

At that particular point in time a discussion of the virtues of field ornithology was futile, so I left it until later. But even when we did discuss it I could tell it was a waste of time. However, it was impossible to dislike him; he was excellent around the camp and his innocent, friendly manner was beguiling.

* * *

This first season progressed reasonably well and our regular routine meant that we accumulated data at a rapid rate. Living conditions improved once our prefabricated cabin (5 m × 3 m) arrived from Newfoundland. Mike Bradstreet was working alongside us by conducting marine surveys of auks around the Gannet Islands. He had arranged for the *Linda-Judy*, which he had chartered for the season, to carry our cabin up from Mary's harbour further up the coast (i.e. south of Grady). As Mike and the crew of the *Linda-Judy* pulled into the Clusters a seal popped its head up in front of them. Mike told us it was all he could do to stop the skipper shooting it. The disturbance caused by a gunshot would have been horrific. As it was I suspected that the auks at the Gannet Islands had been heavily persecuted in the not too distant past. Every time the birds heard the sound of a boat engine they became agitated: Common Guillemots stood up off their eggs and Razorbills emerged from their crevices. The sound of a gun would have caused mass panic. Fortunately on this occasion there was not one.

Living in Labrador, people clearly have to think and act for themselves. I must admit that when our cabin arrived in its numerous sections I would have taken a while to figure out how all the pieces fitted together. Sid however, set to work immediately with tremendous, quiet efficiency and with us acting as assistants. Within 2 days the job was complete, and from that point on we had a warm, dry and well equipped base. We used the cabin for cooking and continued to sleep in tents while the Parcol served as our food and equipment store. We had taken with us all the food we thought we would require for a three-and-a-half month field season except bread. The idea was that someone on the team would bake it. However, in contrast to other camps I had been in, there was no-one with either the inclination or the ability to do this. Fortunately, we made an arrangement with Sid's wife back in Cartwright to bake for us and for the bread to be brought out whenever anyone was passing—which was not very often. Later when Flo came out to stay on the islands for a few weeks we had all the fresh bread we needed. Water was our other slight problem. None of the four smaller Gannet Clusters had any fresh water, and while there was a small stream on GC5 it would have meant going there every few days to collect water. Instead we exploited a seep just by the camp. Water simply dripped out of the peat and trickled down a rock to form a pool 4 × 4 m on the beach. By placing containers under the dripping rocks we had an almost constant water supply. It was fine during that first wet summer, because the rapid rate at which water filtered through the peat ensured that the water at least looked clean. In subsequent, drier summers the water dripped through more slowly and provided what was little more than a concentrated peat solution. It was then that the clear stream on GC5 came into its own.

Once the cabin was complete I really felt that the project was under way. Everyone had become as proficient as they were going to get and knew their duties, so things ran pretty smoothly. There was not much time for relaxation, and with hindsight I suspect some (if not all) of the assistants thought I was a hard taskmaster. After we had completed our observations each day they had to be transcribed neatly into more permanent note books. Having initially suffered the frustration of trying to make sense of my own field notes at the end of a season, I knew how impossible it would be to make sense of other peoples'. We therefore transcribed and partly synthesized our notes every evening. Although this was probably pretty tedious for everyone else, I enjoyed seeing patterns emerge in the data, and this system paid off later when I came to analyse all the results over winter. There was insufficient space on the islands for people to exercise without disturbing breeding birds so our recreation did not amount to much more than the occasional game of frisby on the beach on GC2. On really hot days (and there were several each summer) we skinny-dipped in the pool below the dripping rock, or more bracingly in sun-warmed rock pools. We sometimes ventured into the nearby sea, but this was excruciatingly cold and our brief communal submergence was sheer bravado. Part of my fascination for Labrador was the incongruity of air and sea temperatures. The Gannet Clusters were at the same latitude as my native Leeds in England, resulting in similar warm summer temperatures. The sea, on the other hand, was full of icebergs carried southward in the Arctic waters of the Labrador Current.

* * *

On a murky June day I peered out of the hide window towards my guillemots, but as I did so an unfamiliar movement above the cliffs attracted my attention. On raising my binoculars, all I could see at first were the flailing wings of a Puffin, whose head seemed to be lying at an odd angle. Slowly the image pieced itself together: a blazing yellow eye, huge yellow feet and a helpless Puffin. A Goshawk was rather unsuccessfully trying to pierce the life from a very tenacious Puffin. I could not quite believe it at first. It seems so unlikely

to find a forest-living hawk on these treeless islands. The Goshawk started to rip clumps of feathers from the Puffin's breast, but did not seem to be making any progress in terminating its prey's life. I watched, simultaneously spellbound and appalled. Only rarely does one witness such an event so closely; this one exploded the myth of raptors being swift, efficient killers.

After 10 minutes of half-hearted effort the Puffin miraculously struggled free, and flew off. The Goshawk was immature, and still learning his trade. He was deprived of a meal, but I was relieved for the Puffin.

Over the next few weeks the Goshawk appeared sporadically on the islands, and then disappeared. I doubt if it had died of starvation and I like to think it made its way back to the forests on the mainland.

One of the true pleasures of conducting field work in remote places is that you never know what will turn up next. Making routine observations from a hide increases the chances of obtaining extremely close views of animals that one might otherwise see only at a distance. One spring on my very first visit to the hide, I gently slid open the window to see a white phase Gyrfalcon sitting just 5 m away on the cliff-top. This was a 'lifer' for me, and I was delighted. As it flew off I stepped out of the hide to watch it fly across the water to GC1 where it floated lazily above the Common Guillemots, causing pandemonium among the birds. I wondered whether it was doing this deliberately, trying to flush out an individual with a defect, just as African hunting dogs and spotted hyenas do. The Gyrfalcon eventually flew off over GC2, mobbed and harried by the Great Black-backed Gulls.

* * *

The resident land birds on the Gannets were few and far between. On each island there was a single pair of White Crowned Sparrows, and those on GC2 became very tame, hopping into the cabin for crumbs and becoming everyone's favourites. I never got tired of their simple, repetitive song, which started at first light each morning. Also on GC2 a pair of Shore Larks (Horned Larks) nested in the crowberry scrub not far from the camp. There were also American Pipits, although we never found a nest. On GC5 there was a single breeding pair of Ravens that we heard and saw regularly. Among the non-breeding birds that remained throughout most of the summer there were beautiful Harlequin ducks, which sat on the rocks by the sea beneath our guillemot study areas. There were also a handful of Iceland Gulls, the occasional Glaucous Gull, and very occasionally a pair of Herring Gulls visited the islands.

GC3 was the breeding area for several hundred pairs of Common Eiders. They nested under the *Salix* bushes, and on the few occasions that we visited GC3 we tried hard not to disturb them. Large numbers of loafing Eiders sat on the sea or on rocks just off the south side of GC3 all season. One spring a solitary and striking male King Eider spent several days among the Common Eiders, and enlivened our visits to the 'biffy'. As at Coburg Island (see Chapter 3), the toilet arrangements on GC4 were invigoratingly basic: an open air arrangement on the south side of the island with views of GC5, GC3 and

Grady near the mainland. My first sighting of the King Eider came as I sat gazing over a mass of ice pans: his brilliant orange beak and forehead glowed unusually warm in a monochromatic seascape.

We had several ornithological oddities. The first was a totally albino Puffin which, when I initially saw it flying past in a flock of normal birds, reminded me of nothing less than a cockatoo. It was almost certainly a non-breeding bird because it appeared on many different parts of the Gannet Clusters at different times. We also had a very pale-coloured Common Guillemot whose back, instead of being dark brown, was a beige colour. The most extraordinary feature about this unusual bird was its feet, which were bright orange mottled with black! Its strange colouring apparently did not put off other Common Guillemots because this bird was paired (to a normal mate) and successfully raised a chick.

Another unusual auk we had was a hybrid between a Common Guillemot and Brünnich's Guillemot. We first saw this bird in 1982, and it occupied the same site until our final full season in 1983, but was also seen during a short visit in 1984. Common and Brünnich's Guillemots differ in a number of ways, some of which are fairly subtle: whether the bill tip is light or dark, the shape that the black or brown plumage makes on reaching the neck, and so on. However, its features were exactly intermediate between the two species. The two characteristics that most convincingly made it a hybrid were the fact it had a partial bridle behind each eye (only Common Guillemots are ever bridled), and that it was paired to a Brünnich's Guillemot.

Sue Johnson was the first to notice this odd bird, and we watched it on and off for many days trying to decide whether it really was a hybrid. We then decided to perform a simple test when two new observers, who had both had experience with the two guillemot species, arrived in camp. We 'innocently' asked each of them to count independently the number of Common and Brünnich's Guillemots on several ledges, including the one on which the hybrid had its site. One observer counted it as a Common Guillemot, the other one thought it was a Brünnich's Guillemot. However, when we pointed the bird out to them they agreed that on closer scrutiny it was impossible to classify it as one or the other. These observers were so irritated by our subterfuge and the outcome that they had better remain nameless.

There are actually a few other records of possible guillemot hybrids, so ours was not quite unique. However, it is interesting to speculate about the circumstances in which one species might copulate with, and fertilize the other. I think there are two main possibilities. The first is what I will call the over-enthusiastic copulator hypothesis, in which one species mistakenly mounts the other in a moment of fervour. I actually witnessed this once, when a male Common Guillemot (it would have to be—see Chapter 8) mounted a nearby female Brünnich's Guillemot as she solicited her partner. The second possibility is that one of each species form a pair bond, and produce a chick. This is exactly what Tschanz and Wherlin (1968) observed on the island of Vedøy, Norway, although their love child disappeared at about 10 days of age. There is a very simple way that a mixed pair like this might arise: because the

Heavy seas on the Gannet Clusters. This storm in July 1981, which coincided with the hatching of Common Guillemot chicks, made boating tricky.

two species often breed side by side it is plausible that an egg of one species might have been mistakenly incubated by the other. A young Common Guillemot reared by Brünnich's Guillemot parents would probably grow up thinking it was a Brünnich's Guillemot (or *vice versa*), so when the time eventually came to find a mate, it would search, not for its own species, but for the other. We have no idea whether this could occur in guillemots, but in other species where experimental exchanges have been conducted, it certainly occurs (Birkhead *et al.* 1986).

* * *

The best laid plans can sometimes go awry, and our first season was almost a disaster in terms of the quality of information we collected. On the very day I anticipated that the first Common Guillemot chicks would hatch, we awoke to the sight and sound of a huge Atlantic swell roaring between the islands. From being flat and calm the evening before, the sea was now positively ferocious. The effect was unusual since there was no wind, and I presume the swell we witnessed stemmed from a storm far out at sea. The idea of launching Sid's boat was hopeless, and even if we had been able to do so it would have been impossible to land on GC4. Instead we sat on GC2, watching the increasingly

Home from home: the cabin on GC4, with the main camp on GC2 visible in the background.

Inside the cabin on GC4 with everything to hand: kitchen, working space and bunks.

spectacular waves crashing against the cliffs behind the cabin. The sea remained like this for 4 frustrating days, by which time I had virtually written off our season as a failure. In the event it was not, but the feeling of inadequacy and the irritation of less than perfect data remained with me.

During our days of detention on GC2, Eric and I fantasized about having another, smaller cabin on GC4 to avoid this kind of problem in the future. By the time I came back the following year it was there! Eric and Sid together had designed and built the perfect tiny cabin (just 2.5 m square) at the end of that first full season. There was no wasted space; every bit was efficiently used. There were two bunks, a stove, a writing area, a small side window and, the major stroke of genius, a glass door. The other cabin on GC2 had lots of windows, but all of them were too high up to see through while one was sitting down eating or writing. The glass door on GC2 meant that there was always a view (even if it was occasionally only fog) and plenty of light inside the cabin.

I lived on GC4 once the cabin was there. It was a perfect arrangement. The hides were 30 s walk away, yet the location of the cabin meant that we could not disrupt the birds on our study plots. Also, I could see the GC2 camp, and signalled to the others there, by whistling, whenever we needed to talk to each other on the radio. Living in this limited space, and sometimes sharing it with one of the others, I realized that the relatively huge houses we occupied in the other world were totally extravagant.

There was nothing better than sitting on the grass outside the GC4 cabin at the end of a long day, relaxing with my guitar. The sunsets in Labrador were magnificent, and as the sky turned from its rich red to ultramarine, the stars would slowly start to appear. There were no human sounds, just the occasional whirr of wings as a late Puffin hurried out to sea, or the evocative copulation call of a Common Guillemot on the study plot.

I slept alone in the GC4 cabin in a peaceful solitude. One night I awoke in the dark and staggered outside to urinate. My eyes struggled to make out what was happening in the sky: there was a curious, pulsating brightness, and at first I thought I might be hallucinating. As I rubbed the sleep from my eyes I realized it was the aurora borealis, the northern lights. The scale of the effect was overwhelming and virtually indescribable. I was directly beneath the aurora's azimuth. Curtains of pink and white light radiated down, like shafts of sunlight in a cathedral. Constantly flickering, the flames of light were at once elusive and uncanny; no sooner had my eyes focused on one part of the aurora than it would shift to a new plane. The entire sky was filled with wonderful light, and for an hour I sat on the grass, wrapped in my sleeping bag, totally spellbound.

Auroras, in both Northern and Southern Hemispheres (borealis and australis, respectively) are confined to areas centred on the Earth's magnetic poles, and occur between 100 and 400 km above the Earth. They are truly vast in scale, being thousands of kilometres long, hundreds of kilometres high, and a few hundred meters deep. They occur as the result of the interplay between charged particles and the high velocity wind of protons and electrons streaming from the sun. The charged particles, concentrated into beams are focused by magnetic fields on to the upper atmosphere of the Earth which behaves like a fluorescent screen. Auroras are most intense during and immediately after periods of solar restlessness. This is because solar flares have a major effect on the atmosphere, resulting in increased densities of electrons in the ionosphere.

It subsequently became apparent that 1982 was an exceptional year for the aurora, and on most nights there was some kind of display, although I never saw it again as vividly as on that first occasion. Even at dusk there would often appear a snaky green light over the northern horizon. With time the green band would ripple and grow, stretching and wavering up into the sky.

* * *

Coming back to more terrestrial matters, we were taken aback one evening soon after the start of our first season with the GC4 cabin, when a small rodentish shadow scuttled across the doorway. We put some sunflower seeds on the floor just outside the door and waited, motionless. Within minutes a deer mouse appeared, quite unperturbed by our presence, and started to load up his cheek pouches with the seeds. He disappeared beneath the cabin and returned for more. We rarely saw more than one mouse at a time, and they became the GC4 pets, coming every night for their handouts. It was impossible to determine how many mice there were on GC4, at least without catching them, and we had no desire to do that. We knew that there were mice elsewhere on the island because occasionally walking to or from our open air biffy at night, a mouse would streak across the pool of light cast by our flashlight.

We checked all the other islands quite carefully for signs of mice, but found

none. I was at a loss to know why the mice should exist only on GC4. I also suspected that their total population on GC4 was rather limited, after all the entire island was less than four hectares in extent, with only the top grassy plateau providing anything like suitable habitat. How had they got there? They may have been there since the islands were formed, but that seemed unlikely, especially in view of what happened next.

It all started the following spring when we re-established ourselves on GC4 and looked forward to seeing the mice at the door. Surprisingly they never appeared, and it was not until about 2 weeks later that we discovered why. On that morning Sue, who was outside, startled us by shouting that an otter had just disappeared under the cabin. This seemed pretty unlikely, and on closer questioning we managed to extract a description of this new island mammal: a short-tailed weasel, or what in England would have been called a stoat. However, since we were in Canada, it became Willie the weasel. In due course Willie reappeared, and when we remained still he crept cautiously forward to sniff our toes. We did not really mourn the loss of our mice, which we felt he had almost certainly exterminated, since a weasel was so much more exciting. Soon he started to come right into the cabin while we were working in the evening, and with lightning speed he would run over our feet, our cans of food and across the sleeping bags, disappearing and reappearing with amazing rapidity.

All went well until one morning as I stepped behind the cabin for an early morning pee. I stared down: in the grass lay a large white egg. Initially I thought it might be a hen's egg left-over from someone's lunch. Then I picked it up and saw that it was a Puffin egg, and that it was empty. I was taken aback; we had been checking our study Puffin burrows on GC2 for days and had not yet found a single egg. Clearly Willie had beaten us to it. I placed the eggshell back on the grass beside the cabin wall and was about to go back inside when the weasel appeared from beneath the cabin and started to pull the shell into

Willie the weasel (stoat) taking a Puffin egg under the GC4 cabin (top). Willie inside the cabin (lower).

his lair. A miniature tug of war ensued as I pulled the egg away from him, and he tried to get it underneath the cabin.

The next day there were three empty Puffin eggs behind the cabin. We held a council of war. We were here to study, but also to protect the seabirds, and we could hardly tolerate the presence of an alien predator. Indeed, seabirds choose to breed on islands precisely because there are no terrestrial predators. A wide stretch of water is usually a good deterrent for most predators, but in the Arctic wide stretches of water become a wide stretch of ice for almost half the year. The nearest point from which Willie could have reached the Clusters was Grady, some 11 km of ice-travel away.

The team were divided over what we should do. Most of us thought that

Willie would have to be trapped and 'dealt with', but that proposal met with an angry response from Sue, so we next considered trapping him and releasing him on the mainland. That seemed like a good compromise, and Richard thought that there would be no problem baiting a trap to capture Willie. I thought that a special trip to the mainland to release the weasel was an extravagance we could barely afford, but went along with it. 'Bacon', Richard said 'was irresistible to weasels'. Willie did not find bacon irresistible: he preferred Puffin eggs, and he proved to be impossible to catch. While various, more final, alternatives were being discussed I made a few calculations. There were about 10,000 pairs of Puffins on GC4, so even if Willie ate three eggs per day for the whole time Puffin eggs were available (about 60 days) he could have no more than a negligible effect on the Puffin population. We left him in peace. I must admit that had he taken to eating guillemot eggs things might have been rather different.

I suspect that Willie came to sticky end anyway. As the summer progressed we saw him less and less frequently. We could tell when he was around because his presence triggered a frenzy of mobbing behaviour in our resident White-crowned Sparrows. I believe he ate their babies, as well as lots of nestling Puffins. His period of feasting must eventually have been replaced by one of fasting. All the seabirds leave the Gannet Islands by October, but the sea-ice does not start to reform until December. Although there are some records of short-tailed weasels caching food, it is not known whether they are long-term

Razorbill.

Atlantic Puffin.

food hoarders, so it is possible that Willie starved to death before he had any opportunity to leave—anyway, he wasn't there the following year.

There were some interesting lessons from this experience. Chatting to Sid and his brother Bill back in Cartwright, I learned that it was not unusual to encounter weasels and mice on the sea-ice during the winter a long way offshore. That is presumably how the mice got to GC4 initially. They may well have reached the other islands in the Clusters, but failed to establish themselves there. Willie too, may have visited the other Clusters before settling on GC4 where there were plenty of potential rodent-burgers. The fact that Willie exterminated the GC4 mice, and then succumbed himself, exemplifies the vulnerability of island populations of mammals.

* * *

The Gannet Islands are unique in several respects, not least that they support so many different auk species. There are just six living auks in the North Atlantic and five of them breed at the Gannet Islands: the Razorbill, Common Guillemot, Brünnich's Guillemot, Black Guillemot and Atlantic Puffin. The missing species, the Little Auk, breeds further north but only in west Greenland. However, it was not entirely absent from the Gannet Clusters, once or twice each year a solitary individual would turn up in the middle of the breeding season. I suspect they were young, non-breeding birds, lured in by the sight and smell of all the other birds. My few previous sightings of Little Auks had been of minute specks flying low over the North Sea off the east coast of England. My first genuine view came in the summer of 1982 when Eric pointed out a tiny bird flying amongst a great wheel of Puffins. At first I thought it was a swallow, it was so small and of such different proportions to

Brünnich's Guillemot

the Puffins. It flew above and between the islands for several minutes before obligingly deciding to land on the wooden 'biffy' on GC2. We were on GC4, but still managed a clear view of the bird. It was intensely inquisitive and craned its neck from side to side as though trying to check out the nearby Puffins. Then, it launched itself into the sky, and bee-like, it flew off to the north.

There is a mere handful of places in the North Atlantic where all six auk species breed, one being Grimsey Island off northern Iceland. Slightly more locations, including the Gannet Clusters, hold five species, and more still contain four or fewer species. Interestingly, most auk colonies in the high Arctic contain only a single species: Brünnich's Guillemot (see Chapters 2 and 3).

Although many ecologists tend to study just a single species, few, if any species exist independently of others. The presence of one species in an area can exert a considerable influence on another. The most obvious example involves predators and their prey. Where, when and how a prey species reproduces can be greatly influenced by the presence of predators. Other types of interaction between different species can also be important, if rather more subtle. For example, two species may compete for the same type of food or breeding place. The species most likely to compete are those most similar in their ecological needs, and these are often closely related species. For example, a bird community may include herons, vultures, finches and hummingbirds. It is clearly unlikely that herons and hummingbirds will have much overlap in their ecological requirements, but two species of hummingbird are both likely to need nectar and possibly similar nest sites. The manner in which closely related species within a community divide up resources has been a challenging question for ecologists.

The same type of argument could be applied to auks. They form the most important part of the Gannet Clusters' seabird community, both by being the largest taxonomic group and by being the most abundant birds. A study of the ecological relationships between them therefore seemed appropriate, especially as we were in the unique position of looking at five species simultaneously. In theory, the more species there are within a particular community, the more clearly defined their ecological niches ought to be.

Many years ago the Russian biologist Gause performed some ingeniously simple ecological experiments on single-celled organisms. His experiments led eventually to the formulation of what became known as Gause's principle; that two species cannot coexist if they share exactly the same ecological niche. What this means is that a group of species, such as the auks on the Gannet Clusters, will all have different niches. Each species will utilize different habitats for breeding, and will exploit the food supplies in different ways. The exact manner in which this separation is achieved is what interests ecologists and it was one aspect of our study at the Gannet Clusters that we wanted to investigate.

The auk community at the Gannet Clusters, together with the other seabirds, is shown in Fig. 17. It is clear that two species dominate this assemblage of species, the Common Guillemot and Atlantic Puffin. In terms of their breeding habitat the different species seemed to utilize fairly distinct areas. Puffins nested almost entirely in earth burrows, Common Guillemots used flat rocky areas at sea level but also bred on the few cliffs, as did Brünnich's Guillemots. The Razorbills bred under boulders, in crevices, but also out in the open on rocky ledges. Black Guillemots, of which we found relatively few nests, occupied crevices under very large boulders, and generally kept themselves apart from the other auk species. Of our five species, only the two guillemot species appeared to show much overlap in their breeding habitat, and only where they bred on cliff ledges.

From previous studies conducted elsewhere (e.g. Bertram and Lack 1933) I knew that where the geographic ranges of Common and Brünnich's Guillemots overlapped, they often bred side by side on the same ledge. Several ornithologists had considered these two species to be potential competitors, but none of them had tried to test this idea by making quantitative observa-

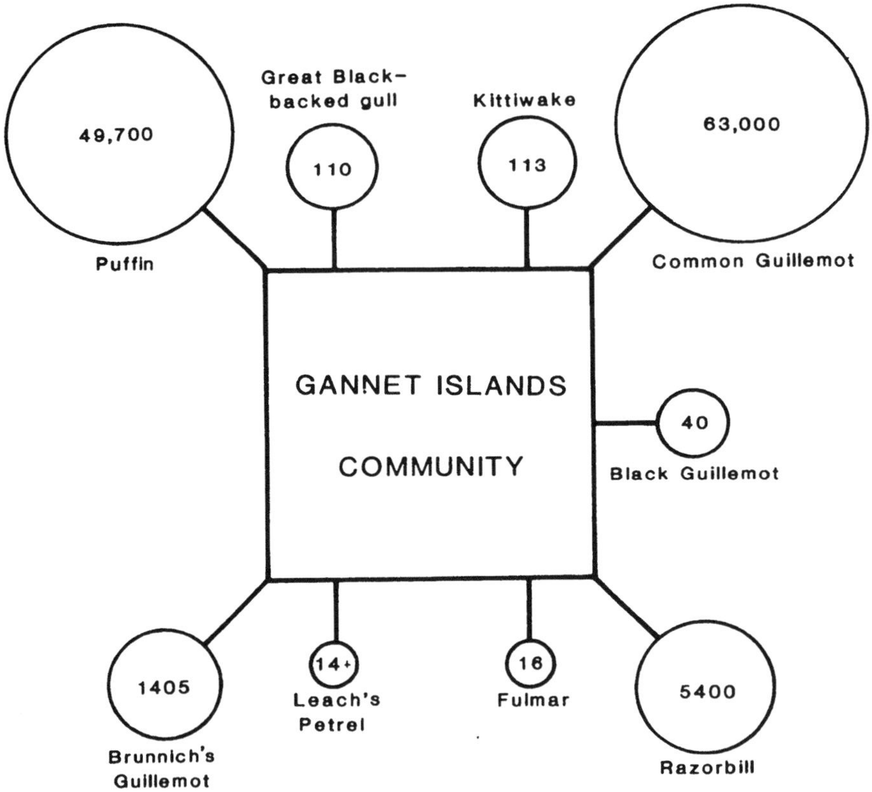

Fig. 17 *The Gannet Islands (Gannet Clusters and Outer Gannet combined) seabird community. The numbers in the circles indicate the number of pairs of each species. From Birkhead and Nettleship (1987a).*

tions (e.g. Belopol'skii 1957; Sergeant 1951; Williams 1974). My objective was therefore to assess in a detailed way the extent of overlap in breeding habitat between the two guillemots and to try to determine whether they competed for breeding sites. At first sight it might seem relatively straightforward to discover whether two species compete for a particular resource, especially something as obvious as a breeding site on a cliff ledge. You might expect to see members of the two species fighting for possession of a ledge, with one species usurping the other. If only it was that simple. In fact, competition between species can take a number of different forms, some as overt as fighting for breeding sites, but others being much more covert. As an example, two or more species may feed on the same species of fish, but by doing so at different times of day they may never meet, yet the consumption by one species will certainly reduce the amount of food available to the other. If this occurs, then both species will suffer in the long run. Indeed, the ecologist's definition of competition is that it results in a negative effect on all parties.

* * *

The area in which I decided to investigate the two guillemot species was a sheltered cove on the north side of GC4. This small bay contained several thousand pairs of Common Guillemots and a few hundred pairs of Brünnich's Guillemots, and this was where I spent most of my time. My dual aim was to make a detailed, numerical assessment of the type of breeding site used by each species and to find out how the type of site influenced the breeding success of each species.

Both Common and Brünnich's Guillemots use a patch of rock ledge about the size of your hand for breeding, and they very often return to this exact site year after year. I used the scheme devised by Tony Gaston (Chapter 6) to quantify both the physical and social features of each site. The physical characteristics comprised the width of the ledge on which the site occurred, whether the site had one or more adjacent vertical walls and whether it was level or sloping, and, if it sloped, whether towards the sea or the cliff. The social aspect simply recorded the number of neighbouring guillemots adjacent to the site. This sounds complicated, but after doing a dozen or so, this scheme was very simple to use.

The other part of the study was one which took considerably longer, and comprised our recording the breeding success of each pair of guillemots. The final objective was then to combine this information so that I could tell first, whether the characteristics of the site played any part in determining breeding

Two of our Guillemot hides on GC4, June 1982.

success, and second whether the quality of the sites used by the two species differed, and hence whether the two species were in competition.

* * *

Sitting in a hide opposite a cliff full of guillemots with a view to recording what proportion of pairs successfully rear a chick can look like an impossible task. Earlier attempts to measure the breeding success of both guillemot species involved climbing onto the breeding ledges, writing a number on each egg and on subsequent visits seeing how many eggs had disappeared or whether new ones had been laid. I suppose that the early students of breeding success thought that the tenacity with which guillemots incubate their eggs or brood their chicks meant that they were immune or indifferent to this kind of disturbance. They are not, and in addition this method, which is dangerous for the biologist, causes extensive egg and chick mortality, and so fails to provide a meaningful estimate of breeding success.

The technique we used was one I had developed during my PhD on Skomer Island, and involved taking a black-and-white photograph of the chosen study group during the middle of the breeding season. On the photograph all the individual sites are numbered, preferably in a logical sequence, such as left to right or top to bottom, since this greatly increases one's ability to recall the site numbers. Each site is then checked every day through binoculars or a telescope, usually from a hide, to record the presence or absence of eggs or chicks. The secret of efficient recording is to memorize the site numbers, sometimes several hundred of them, so that one maximized the amount of time watching the birds rather than looking up site numbers on the photographs. It was not until I returned to Skomer Island after an interval of several years, with Ben Hatchwell, a new research student, that I realized how deeply branded into my brain the site numbers had been. As I went through the study plot with him, my site numbers from 10 years earlier came back with alarming alacrity. The thought of all those brain cells tied up with so much junk filled me with horror.

My non-invasive (but mind scarring) technique was extremely time-consuming, but also very rewarding. We got to know certain individually recognizable birds quite well, and we ended up with as precise a measure of breeding success as it is probably possible to get. Our 1.2 m square plywood hides became like a second home for each of us. It was important that they were comfortable because we were all going to spend long periods of time in these hides, 6 to 8 hours a day on average, and sometimes as long as sixteen hours. They were draught- and water-proof, with cushioned seats, a flat, level wooden floor, and each hide even had a sliding perspex window for when it rained. This kind of field comfort is all relative, for once you have such a set-up, it is easy to think of further (inaccessible or inappropriate) luxuries one would like.

* * *

There was one aspect of comfort over which we had little control. This was the appearance of unwelcome visitors, albeit rather small ones, called *Ixodes uriae*!

A 14-day-old Puffin chick badly infected by larval and nymphal ticks. This bird had several hundred ticks around its eyes and had lost much of its down as a result.

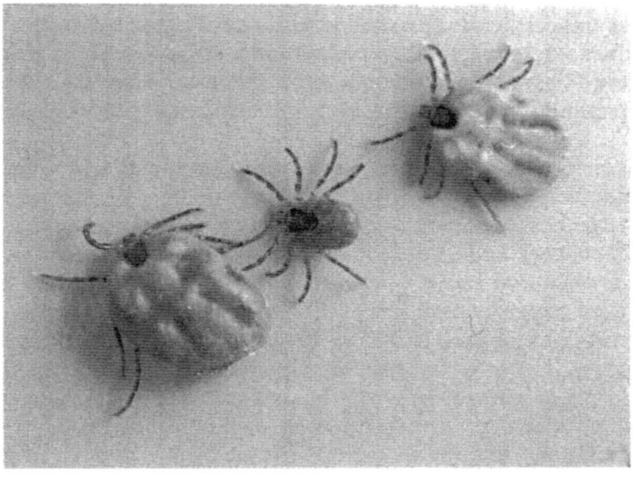

Ixodes uriae: *ticks. The central tick is probably in its second summer, and the outside ones in their third summer. (Photo: D. Hollingworth).*

Attracted by the aroma of our bodies, these seabird ticks made their way into our hides in large numbers. The ticks were mostly young 'larvae', like flattened spiders and about the size of a pin-head, although we sometimes encountered larger ones. They clambered around the hide, hoping (if ticks 'hope') eventually to gain entrance to the bottom a trouser leg. From there they would slowly ascend to any warm crevice and insert their sucking mouthparts.

We all developed heightened sensitivity to any kind of inner-trouser movement, since way-laying one of these little darlings *en route* was far more preferable than the alternative. Although ticks are nobody's favourite, they certainly have their favourites, and fortunately I was not one of them. In five visits to the Gannet Islands I was never bitten, but some of my colleagues attracted more than their fair share of these repulsive and potentially dangerous beasts.

Once a tick has embedded its head inside you it simply gorges on blood. This is completely painless. After a few days the tick is engorged and it drops off, but it is then that you begin to suffer. Substances in the tick's saliva are extremely irritant, and the bite can itch uncomfortably and sometimes intolerably for up to 3 weeks. In addition, there was always the risk the tick might transmit some unpleasant disease. Certain ticks are infamous for transmitting diseases such as Rocky Mountain Spotted Fever and Lyme's Disease, and bird ticks are also known to harbour potentially hazardous viruses (Chastel 1988). It was presumably because of this that the American airforce funded a great deal of tick research at one time, probably with a view to conducting biological warfare on hostile powers.

People who discovered an attached tick usually asked someone else to remove it discreetly for them. Ticks were sufficiently unpopular that each of us kept a tick gibbet in our hides and we compared scores at the end of each day.

I could just about tolerate ticks in the hide, but I hated them in my sleeping bag. Sometimes we inadvertently carried ticks on our clothes, which subsequently found their way into a sleeping bag. There was no worse sensation than feeling the creeping motion of a tick on your body just as you were dropping off to sleep.

The larval ticks that plagued us were miniature adults which had hatched from eggs laid the previous summer. Under normal circumstances these tiny ticks would attach themselves to a bird host, usually a guillemot or Puffin, feed for a about a week and then drop off to moult and hide in the soil until the next summer when the process would be repeated. Not until after 4 years would the tick emerge as an adult. Fully grown ticks are loathsome beasts, about twice as big as a baked bean and battleship grey in colour. They took a lot of killing too. This is not surprising since they have evolved to withstand the various forms of punishment that hosts and potential hosts would try to inflict on them. They had to be trodden on with one's full weight before they expired with a satisfying 'pop'. An adult tick has two priorities: to feed and to mate, in that order. Only once they had fed and were fully engorged would the females copulate. Interestingly, just like their Common Guillemot hosts (see Chapter 8), females sometimes copulate with more than one male. Mating takes about 24 hours and afterwards the male drops off and dies almost immediately. The female, on the other hand, lives a further 3 or 4 months as she lays her clutch of 500 eggs to start the cycle again (Eveleigh and Threlfall 1974; Steele *et al.* 1990).

Although much is known about the biology of *Ixodes uriae*, we still know very little about the effect they have on their seabird hosts.

* * *

To be able to obtain an accurate measure of the birds' breeding success we had to start making observations early in the season before any birds had laid. Sites were checked every day to establish the exact date on which each female laid her egg. This was one of the most interesting periods of the birds' breeding cycle, especially with respect to the Common Guillemot's remarkable mating antics (see Chapter 8), but also in terms of their egg-laying behaviour. On Skomer Island I had seen the odd egg actually being laid, but here, because our hides were so much closer to the birds, and we spent such long periods of time just watching, we saw a high proportion of eggs being laid. The first signs were obvious: the female stood completely upright on her legs, and held her wings slightly outstretched away from her body. Often her head was held vertically with the beak gaping open as she struggled for breath. Every so often she would strain downwards and flop forward exhausted onto her belly. Then, perhaps on the next go, you could see the first part of the egg protruding from the cloaca. With an almighty heave, the entire egg would emerge in a fraction of a second. Observers and, I presume, the bird itself, always sighed with relief once the egg was laid. The pointed end emerges first and the Common Guillemot's short, stiff tail guides the egg forward so that it comes to rest between the female's legs. At that point in time the egg is at its brightest and cleanest, and the colours range from deep turquoise, to bright green, white and occasionally dark brown. The female is intensely interested in her achievement but after a few minutes parental pride and memorizing the egg's features, she settles down to incubate.

For the next 32 days she and her partner take it in turns to incubate for about 17 h at a time (range 1–38 h) until the egg hatches. Most frequently the egg is incubated on and between the feet, with the pointed end directed inwards (70% of cases), less often directed outwards (20%) and in the remaining 10% the egg is held at 90° to the bird's body. Inevitably some eggs are lost during incubation, from a variety of causes. Even after the chick has hatched the parents continue to take turns in guarding it from predators and other dangers at the nest site. At this stage though, the shifts are much shorter, an average 12 h (range 7–20 h) if we include the overnight period, but just 4 h (range 1–14 h) for only the daylight period. Each partner takes an equal share in these incubation or brooding duties while the other bird forages away from the colony. By about 21 days of age the chick is ready to 'fledge', and once it had gone we could score the breeding attempt as successful.

Of course, not all eggs hatch and not all chicks fledge. Some are lost to predators, others simply fall, or are knocked from their breeding ledge by clumsy neighbours. On our study plots in Labrador, 82% of Common Guillemot pairs successfully reared a chick to fledging, whereas only 63% of Brünnich's Guillemots did so. This difference between the two species occurred in each of the 3 years in which we measured breeding success.

* * *

It was not until I was back in Sheffield that, with the help of a statistician colleague, John Biggins, I analysed the information on the breeding site

Common habitat. On this cliff study plot Common and Brünnich's Guillemots bred side by side. At the top and right hand side the dense aggregations are Common Guillemots. Our hybrid guillemot (see text) bred in the closest group on the far right of the picture.

characteristics of the two species. Even before we had analysed our data I was pretty sure there would be a difference between the two species, because just looking at the ledges it was apparent that Brünnich's Guillemots occupied narrower ones than Common Guillemots, but we needed to attach numbers to the differences and to apply the necessary statistics to be sure.

The site characteristics of the two species did indeed differ in the way I had anticipated and the most frequent type of site used by Common Guillemots was a broad ledge, either level or sloping and with no rock walls. In contrast, Brünnich's Guillemots most often used narrow ledges that were level and had one rock wall. In fact almost all Brünnich's Guillemots used sites that had at least one wall. Judging from the way this species 'sits' while it is incubating, it is probably most comfortable leaning its breast against something.

Next, I looked at the information for each species separately to see if the type of site a pair occupied had any bearing on their likelihood of successfully rearing a chick. Our analyses showed that all the site features we recorded were important in this respect, when considered separately. Taking Common Guillemots first, we found that those birds breeding on the broadest ledges were the most successful. Birds on level ledges did better than those on sloping ledges, and birds also did better the greater the number of close neighbours they had. Of course, not all the site features were independent of each other; for example, a site on a very broad ledge was much less likely to have a wall than a site on a narrow ledge. We were able to disentangle some of these effects statistically, and this showed that the most successful birds were those breeding on broad ledges with a wall.

It was somewhat unexpected to find that almost exactly the same features were important in determining the breeding success of Brünnich's Guillemots. This was surprising because, in contrast to the Common Guillemot, there was a large discrepancy between the sites which Brünnich's most commonly used and those which were most productive. For Brünnich's Guillemot, just as in the Common Guillemot, breeding success was highest on the broadest ledges. Most Common Guillemots occupied broad ledges, but most Brünnich's Guillemots occupied narrow ledges, hence their relatively poor success. Thus, most Common Guillemots bred on the best sites, but only a small proportion of Brünnich's Guillemots bred on good sites. This result strongly suggests the existence of competition between the two species and indicates that if there had been no Common Guillemots at the Gannet Clusters the Brünnich's would have been free to occupy the broader ledges and as a result would have been more successful in rearing offspring (Fig. 18).

The competition for breeding sites between the two guillemot species is fairly subtle. There was no indication that fights between them were any more frequent than between members of the same species. Nor was breeding success affected by whether a bird had neighbours of the same or a different guillemot species. One reason why Brünnich's Guillemots came off worse in this relationship was that the total amount of breeding habitat available to this species was rather limited. Brünnich's Guillemot breeds only on cliffs more than 5 m high. The Common Guillemot, on the other hand, is much more flexible and can breed on cliff ledges, but also on broad, flat areas close to sea level. Consequently, Common Guillemots at the Gannet Islands were able to breed anywhere that Brünnich's could breed, but the Brünnich's were confined to the relatively small amount of cliff habitat. What is more, because of their greater flexibility Common Guillemots had a much greater amount of suitable breeding habitat available to them, so most of them could avoid breeding on narrow ledges.

Considering the Gannet Islands and Outer Gannet together there were about 40 times as many Common as Brünnich's Guillemots. It is not clear whether the latter's small population size is a direct result of their limited cliff habitat, or of some other factor. But whatever the explanation, this difference in relative abundance means that Common Guillemots exert much more pressure, in terms of say, breeding space, on Brünnich's, than *vice versa*. Thus,

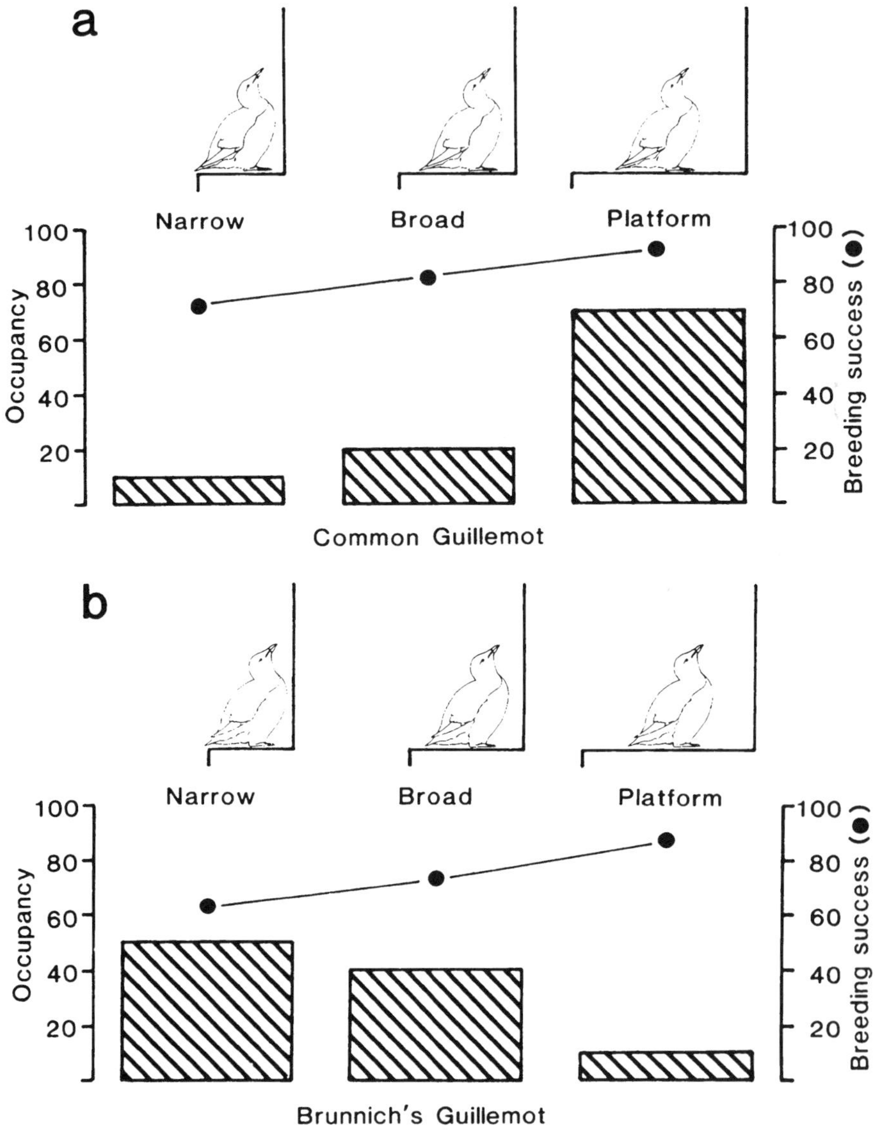

Fig. 18 *Interspecific competition between Common and Brünnich's Guillemots. The figure shows the percentage of ledges of three different width categories (narrow, broad and platform) occupied by each species and their breeding success on those ledges. (a) Most Common Guillemots breed on platforms, fewer on broad ledges and very few on narrow ledges and breeding success is positively associated with ledge width (i.e. the broader the ledge, the more likely the birds are to rear a chick). (b) In contrast, most Brünnich's Guillemots breed on narrow ledges with relatively few on platforms. Yet (as in Common Guillemots), breeding success is greatest on the broadest ledges. The difference in the pattern of ledge occupancy and breeding success between the two species indicates that Common Guillemots are competitively superior and keep Brünnich's Guillemots off the most productive sites.*

while our results provide good circumstantial evidence for the existence of competition for breeding space between the two guillemots, the competition is asymmetric, with Brünnich's Guillemot getting the poorer deal. To put it another way, Brünnich's Guillemots would probably have been delighted if all the Common Guillemots disappeared, but I doubt if many Common Guillemots would have even noticed if the Brünnich's disappeared.

* * *

Our camps in Labrador contained as many as seven people at a time, and over the 3 years about 20 different individuals were involved in the study. The atmosphere in camp was overwhelmingly friendly and positive, but I suppose it would have been unrealistic to expect everyone to get on with everyone else all the time. Small communities are well known for distorting social relationships, and trivial problems can sometimes take on gigantic proportions. In extreme cases, individuals that deviate too far from what is regarded as normal behaviour are said to be 'bushed'. Erick Greene's team experienced this an unpleasant way in 1979 (Chapter 3), when just before leaving the Cape Hay camp, one of the mammal watchers, in an act of unparalleled magnanimity, defecated into their 45 gallon drum of drinking water.

It was during the difficult moments when personal relationships were tense that I was reminded of some advice T. G. Longstaff gave David Lack in the 1930s when Lack was planning his expedition to Bear Island: 'the ideal number for an expedition is one, and the party should consist of the minimum number above that to attain one's ends'.

The problems sometimes associated with small communities can be avoided if individuals are bonded together in a common cause. I have noticed that if there is a scapegoat, especially if it is someone outside the camp, this can often help to maintain the internal structure of a camp. Irrespective of problems and scapegoats one of the aspects of camp life that I have enjoyed most is the close friendships that can develop out of common interests and a strong interdependence upon each other. My position as leader inevitably meant that my view of the social relationships in camp differed from those of the others, and I am sure there were plenty of things that escaped my attention. Obviously there were benefits associated with being in charge, but there were also disadvantages, one of which was dealing with relationships when things went wrong.

I have fond memories of virtually all the people that helped us over the years and we were extraordinarily lucky in the combinations of personalities and talents we had in our camps. I have only one bad memory. Usually when I met members of the team, most of whom were skilfully chosen by David Nettleship, it was apparent straightaway that they would fit into the camp situation. However on one particular meeting it was impossible to make eye-contact, and I experienced a twinge of unease. It is always difficult to distinguish between cause and effect in these instances, but my fears for this person were eventually realized in full. This individual—let us call him Dick—was too much of a loner to be part of a team, but by the time we met it was too late since we were already on the islands. The unfortunate fact is that a

Habitat differences between the two Guillemot species. Brünnich's Guillemots bred only on the cliffs (top), and Common Guillemots bred predominantly on the broader areas at the foot of the cliffs.

single unhappy individual can act up like an ulcer and irritate everyone else in the camp. In this particular case a succession of irksome incidents eventually developed into a dangerous one.

The constant use of our two Zodiacs meant that there were always minor rips on the hull to repair and outboard motors to be coaxed back to life. On this particular occasion, Dick, who fancied himself as something of a mechanic, offered to grapple with a particularly irascible motor. Aware of his general cavalier attitude to instructions and the dangers associated with boating, I firmly suggested that he stayed close inshore to GC2, just in case he was unable to restart the motor. The rest of us went inside the GC2 cabin to write up our field notes. Just 15 minutes later I stood up to look out of the window to discover a thick fog and no sight nor sound of Dick on the water. I ran down to the mooring place and shouted and tried to penetrate the mist with my binoculars, but to no avail. Back in the cabin Dick's one ally lost his cool and accused everyone else of being irresponsible. Fortunately, the others were much more practical: Sue called the coastguard in Cartwright and started to prepare some emergency supplies. We made a rapid repair job on our other Zodiac, and Eric and I got ourselves ready to start looking. As the various preparations were being made the fog lifted slightly above sea level, allowing us to scan the surrounding area, but there was no sign of Dick. I had the most awful sinking feeling in my stomach and hoped that he might have been able to get to one of the other islands.

Our search started in a mood of stony irritation liberally mingled with dread. We set off towards GC6, all the while searching the horizon. Half-way between GC2 and GC6, I saw an orange dot about 15 km distant; it resembled the boating jacket which Dick had been wearing. We scrutinized the blob of colour as it appeared and disappeared between the distant waves, but after some initial optimism, we decided that it was a large fishing float and not a person. We reached the most distant of the Gannet Clusters, and starting at the north end, slowly made our way round the coast. Eventually, and with a sigh of relief, we saw the Zodiac pulled up on the landward shore. When we met him, Dick was his usual cool and surly self, unable to appreciate what all the fuss was about. . .

Perhaps predictably, our relationship remained poor and eventually declined until it reached a state of mutual resignation. Dick continued to go through the motions of making observations, but during our communal sessions during which we transcribed field notes I started to become concerned about his results. More and more inconsistencies began to appear in his data, until in the end I decided simply to bin everything he had done on his own. There is nothing more unsettling in science than someone you cannot trust. Fortunately, his responsibilities comprised an expendable part of the study, and did not involve guillemots.

* * *

The second question concerning the Gannet Clusters auk community, was whether the five species competed for what they fed to their young. The chick-rearing period is that part of the season when the energetic demands of every individual, and hence the entire seabird population is at its maximum. The adults are working hard flying to and from the colony ferrying food to their hungry chicks, and these extra mouths need to be fed. Since chick-rearing might be the most energetically costly part of the breeding cycle we predicted that the most efficient system would be one in which each species exploited a different type of prey. In that way the different auk species would avoid competing with each other. It is important to note that this would have come about through natural selection operating on individuals (not species). For example, any particular Razorbill that decided to eat the same food as the Common Guillemots would be at a disadvantage relative to those Razorbills that exploited some other kind of prey—everything else being equal. In this case, 'everything else' was the relative abundance of the different prey species. If things were not equal and the prey most Razorbills chose to eat was scarce, then they might actually be better off competing with Guillemots.

All this means that natural selection will favour those individuals, of whatever species, that get the balance between these conflicting pressures right. This does not mean that a Razorbill or a Puffin consciously assesses the situation and then decides what prey to go for. The process of natural selection is entirely mechanical: those individuals that do the wrong thing leave fewer descendants; those that get it right leave plenty and the genes for that particular foraging pattern are maintained in the population.

We needed to know what each species fed to its offspring. This sounds relatively straightforward—surely you just go and look at what they are stuffing down their offspring's gullet? Unfortunately, it is not quite so simple. Imagine you are watching a group of Brünnich's Guillemots. You know from spending many hours each day in the hide that all the birds on your ledge have got chicks (you might even have seen a few hatch from the egg), but at this particular moment you cannot see a single chick. This is what it is like most of the time because the parents brood their offspring very tightly, to keep it warm and out of the harsh gaze of the predatory gulls. Suddenly you become aware of an adult bird flying towards the ledge at high speed and landing clumsily amidst its neighbours. As the bird alights it tucks its head in and reaches down towards the back of the ledge where its chick has emerged from under the mate. Then, a second or two later the recently arrived bird stands upright, shakes some fluid from its beak, and settles down to preen. Was there a fish? Was that a feed? Perhaps you got a glimpse of silver sliver held lengthwise in the bill, suggestive of a fish, but as for what species and how big it was: who knows?

This is absolutely typical of the way the Brünnich's Guillemots feed their young. Recording how often and with what the chicks are fed was far from easy. The situation in the other species was similar but marginally easier. With practice and unwavering concentration it was possible to anticipate when a bird was about to come in from the sea and feed its chick. This maximized the time we had for assessing the size and identity of the prey item. However, sometimes, even when we could clearly see a fish in a bird's bill, we had no idea of what it was. When I had conducted similar feeding watches of auks on Skomer the identification of fish was rarely a problem since 99.9% of the 2000 or so fish I saw fed to Common Guillemot chicks were sprats. In Labrador the diversity of prey species was considerably greater, but no-one knew very much about what was actually out there in the sea, so we had to start from first principles and obtain a set of specimens. Although the business of fish identication was frustrating initially, it made conducting feeding watches great fun since you never knew what species was going to turn up next. Obtaining fish specimens was also one of my favourite forms of fieldwork, combining my basic hunting instincts with collecting data! Fish are so aesthetic I simply enjoyed seeing them, and straight from the sea some of them smelt beautiful too. Capelin, for example, are dark green along their back, silver on the flanks, with an iridescent lilac sheen, and smell of sliced cucumber.

Common Guillemots were the easiest of the auks to conduct feeding observations on. The incoming bird, having flown several kilometres, perhaps from Table Bay (see Fig. 12, Chapter 5) would alight on the sea at the edge of the study colony. It then swam the last few metres to hop ashore with its prize held lengthways in its bill. The bird might then take as long as 5 minutes to reach its site and its chick. We thus had plenty of time to try to identify each fish and estimate its size against the length of the adult's beak, which we knew from the birds we caught, to be 42 mm long. In our first season in 1981 we were able identify only the capelin, the rest of the fish were completely unknown to us. Rather than waste the information I simply categorized the fish into a number of clearly recognizable types. One was brown and thin with a whitish, but not silvery belly, while another was orange and snake-like with a large black spot on the dorsal fin.

The only way to find out which species we were dealing with was to get our hands on some. There was no colour field guide to Labrador's little fish, and most keys to fish identification recommended counting fin rays and examining several other esoteric features. I managed to find a few specimens on, or under the birds' breeding ledges, but most of these were either dried up, or damaged and of little use for identification purposes.

This problem does not arise when studying Puffins, since one of the standard study techniques involves netting the birds as they returned to the colony with fish. Once caught, the Puffin drops its beakful of fish and, because Puffins breed on nice grassy slopes, it is easy to collect the dropped fish. In this way one can identify the prey species, weigh and measure each fish and know the weight of the whole meal. This technique is simply not appropriate for other auks, because their cliff sites are so inaccessible, but here on the Gannet Islands relatively few Common Guillemots bred on cliffs. The majority bred on flat rocky ledges, not much above sea level. I wondered whether I could find a place where it was possible to intercept incoming birds and relieve them of their fish, but without frightening those in the adjacent breeding colony. After clambering around the periphery of GC4 I found just such a place on the south side of the island. It was absolutely ideal, I could conceal myself behind a small rock wall, peep over the top every few seconds and when I saw a bird coming in, I could try to net it. The technique I used was a very old one and involved using a net on a long handle. In seabird terminology this is a *fleyg*, the name used by the Faroese auk catchers. For centuries native peoples throughout the Arctic have used the same method to catch small seabirds, such as auklets and the Little Auk.

Initially I was inept; in one swift movement the net had to be lifted from its hidden horizontal position until it was vertical and in the bird's flight path, and at exactly the right moment. If I emerged from my hiding place too early the bird would veer off and be out of reach, and it was easy to be too late. In addition, the long handle of the net was quite heavy and after a few futile attempts in rapid succession the net seemed to get heavier and heavier and I got slower and slower at raising it into its catching position. Once in a while I caught a bird, not always with a fish I admit, but it convinced me that the technique had potential (even if I was uncertain about my own). I felt sorry for the birds I caught because they seemed to hit the net rather hard and it often took me several minutes to extricate them.

I discovered by accident that birds which I merely bumped on the breast with the rim of the net, readily dropped their fish. Thereafter I stopped trying to catch the birds, but concentrated on simply persuading them to release the fish from their bill. This was much less traumatic for the birds and for me, and it speeded up the rate at which I accumulated specimens. I eventually became sufficiently adept that I could bump the bird with the net in one hand and catch the fish as it dropped, with the other.

In this way I rapidly built up a reference collection, and by flicking through Leim and Scott's (1966) *Fishes of the Atlantic Coast of Canada* I was able to identify the majority of species. There was one species however, the 'orange snake', that remained a mystery. The nearest I could get to it in Leim and Scott's book was the fish doctor *Gymnelis viridis*, but there was one feature that simply did not fit. My fish specimens had a very obvious black ellipse on their dorsal fin, and the book said quite categorically that the fish doctor had no black spots on its dorsal fin. I took some specimens home and later sent them to Alwyn Wheeler at the British Museum. He too was puzzled initially, until he looked at the museum's collection of fish doctors. He discovered that the appearance of this species varied considerably through its geographic range, and that those from Labrador, but nowhere else, did indeed possess a dorsal fin with a black spot.

One of the most spectacular fish we saw being fed to chicks occasionally, but never managed to get a specimen of, was the snake blenny. As its name implies, this was a thin species, often much longer than the Common Guillemot chicks they were fed to. It was always amusing to see a chick take

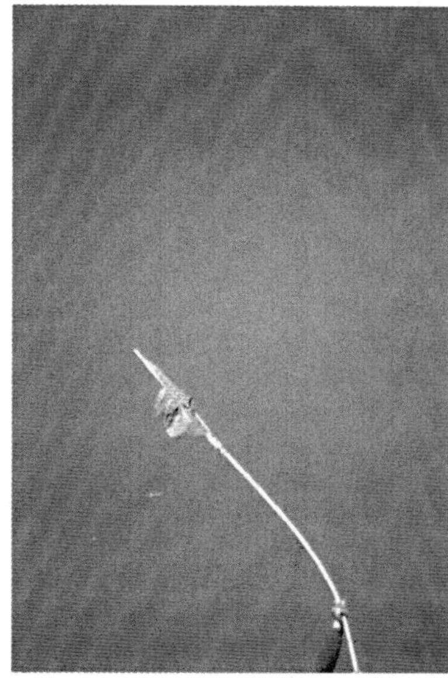

Fleyging *Common Guillemots on GC4 (on some days at least this was hot work – hence naked torso). (a) and (b) Near misses – Common Guillemots with fish avoid the net. (c) A catch, and (d) another catch, and the fish can be seen in mid-air below the net. (Photos: S. Johnson).*

one of these from its parent, swallow it head first, only to have 10 cm or more protruding out of its mouth like a gigantic tongue.

Although we were fascinated by the rarer and more unusual species of fish fed to the Common Guillemot chicks, there was one species upon which the entire community hinged, the capelin. This is such a remarkable species that it is worth a digression.

* * *

I stood on the jetty in Cartwright watching a half-dead fish swimming crookedly towards the surface. Without warning a rapid succession of stones sliced through the water and finished it off. The fish was a spent capelin, and having completed its spawning it was now used by the local boys to help kill time.

The capelin's synchronized spawning is an integral part of Labrador's culture. During his survey of the Labrador coast in 1868, Lieutenant Chimmo observed and described the capelin's reproductive behaviour:

> The manner of the Capelin spawning is one of the most curious circumstances attending its natural history. The male fishes are somewhat larger than the females, and are provided with a sort of ridge projecting on each side of their backbones, similar to the eaves of a house, in which the female is deficient. The female on

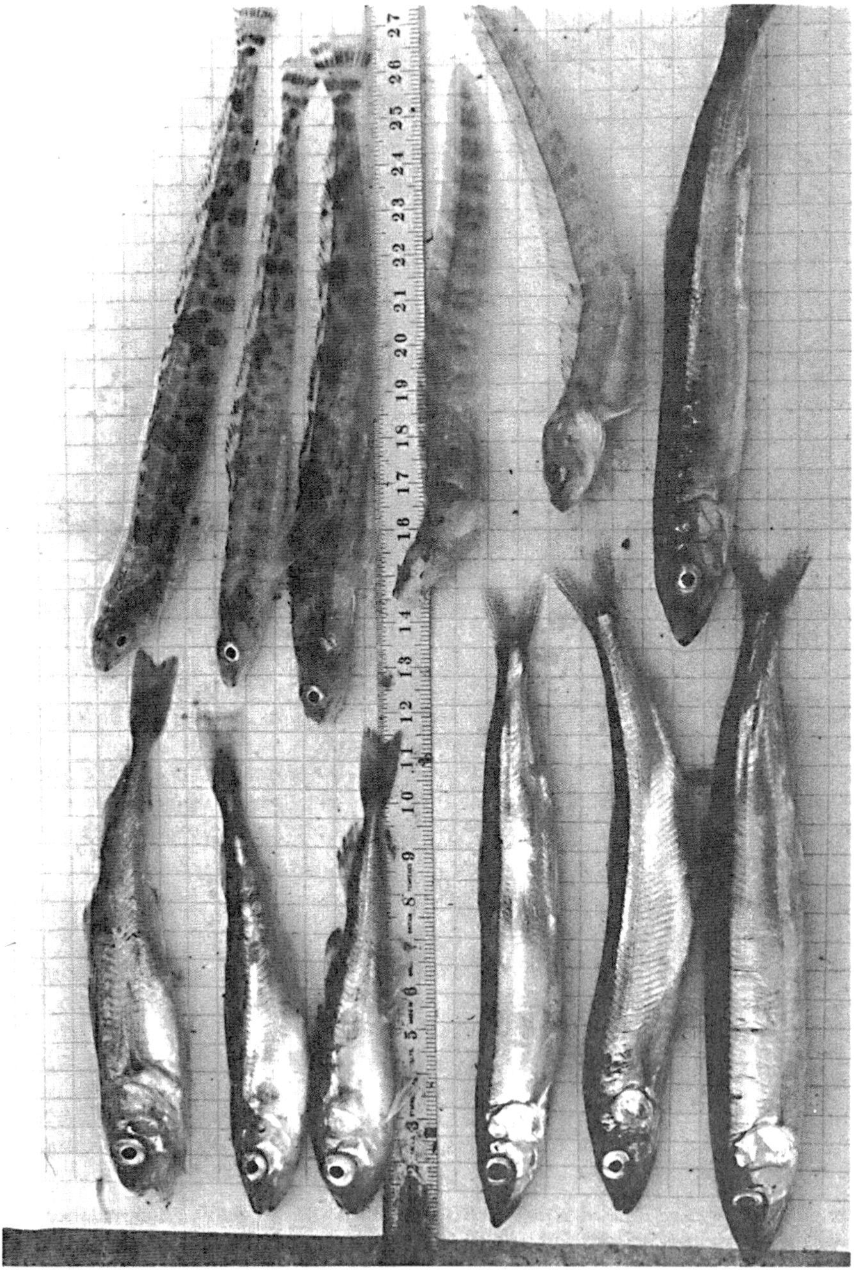

Some of the fish delivered by Common Guillemots to their chicks. Left (top–bottom): 3 arctic cod, 3 capelin. Right: 3 daubed shannies, fish doctor, stout eelblenny and another capelin.

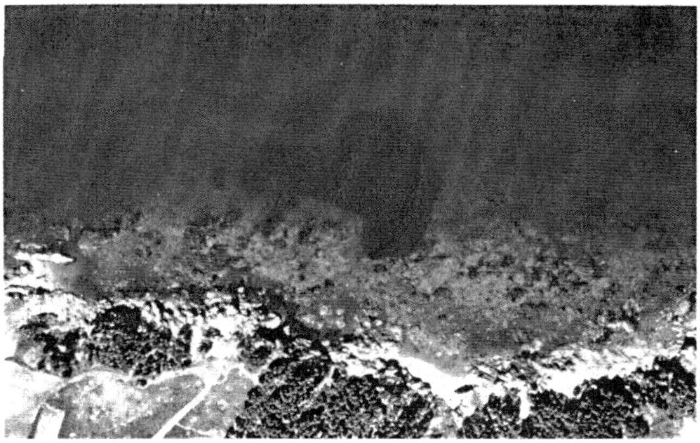

An aerial view of a dense shoal of Capelin lying close to the shore prior to spawning. Some idea of the size of the shoal can be gauged from the road at the bottom left. (Photo: B. S. Nakashima).

> approaching the beach to deposit its spawn, is attended by two male fishes, who huddle the female between them, until her whole body is concealed under the projecting ridges, and only her head is visible. In this state they run all three together with great swiftness upon the sand, when the males by some imperceptibly inherent power, compress the body of the female between their own so as to expel the spawn from an orifice near the tail. Having thus accomplished its delivery the three Capelin separate, and paddling with their whole force through the shallow surf of the beach generally succeed in regaining once more the bosom of the deep.

The capelin spends most of its year offshore, but during June and July it moves inshore to spawn in the surf on sandy or gravelly beaches. Spawning occurs only when the sea temperature has reached 40–47°C, and usually at night in calm seas. Spawning occurs later the further north one travels in Labrador. The spectacular nature of the Capelin's spawning is not apparent from Chimmo's description. Vast, black shoals of Capelin move into the vicinity of the spawning bays, heaving themselves in huge numbers onto the beach. The eggs are golden yellow and each female lays several thousand; the sea becomes milky with the males' sperm. Local people take large quantities of this easily exploited fish, scooping them out of the surf with buckets and using them to feed their dogs, as fertilizer and as bait for cod. More recently Capelin have been exploited for their ova alone, simply to trigger multiple gustatory orgasms for the Japanese (see Chapter 4).

* * *

Discovering what the auks other than guillemots fed to their offspring was relatively easy. Black Guillemots simply allowed us to sit near their breeding sites to see them returning with prey. We used a similar technique with Razorbills, and I also tried to acquire fish from incoming Razorbills at the same time that I was collecting fish from Common Guillemots. One reason why the *fleyg* method worked so well with Common Guillemots was that their manoeuvrability in the air was so limited that even when they saw the net they could not dodge it. The same was not true of the Razorbill. Their ability to avoid my net, a result of their lower wing loading (see Chapter 4) was remarkable, and I started to view them in a new light. Despite the superficial similarity to guillemots in their design, their appearance was deceptive and their aerial agility was such that I managed to relieve only a few Razorbills of their fish.

The Puffin chicks' diet was ascertained by a mixture of observation and netting incoming birds. The two techniques were essential since observation alone gave very biased results. This was because birds carrying relatively small loads of fish were able to fly directly into their burrow, giving the observer no chance to record whether or not there were fish in the Puffin's beak. On the other hand, Puffins carrying two or even three large, conspicuous capelin more or less staggered into their burrow.

After our first frustrating season of trying to determine what Brünnich's Guillemots were feeding their chicks, I decided to build a small hide directly adjacent to group of breeding birds on GC1 to try and get a closer look at the fish. The hide was erected early the next season before any birds had laid, and was positioned so that I could see along the length of the ledge and hence see the chicks of several pairs. The nearest birds were just 1 m away, since the Brünnich's Guillemots were completely unperturbed by my presence in the hide. From my point of view being so close to the birds made me feel as though I was part of the colony, and more importantly it allowed me to identify virtually every fish that was brought in for the chicks. In this way I was able to verify our identifications made from the other, more distant hides on GC4.

With one interesting exception, each species of auk exploited a different species of fish. Black Guillemots fed their chicks mainly on sculpins, Razorbills took mostly sandeels, Brünnich's Guillemots took daubed shannies, while both Common Guillemots and Puffins preyed extensively on capelin (Fig. 19). The situation in the last two species was interesting because these were the most abundant auks at the Gannet Islands (Fig. 17), and I did not expect their chick diets to be so similar.

* * *

Simply because two bird species exploit the same prey species does not necessarily mean they are in competition with each other. If food is superabundant there cannot be competition. Even if food is not overly abundant, competition can be avoided in several different ways, for example, by feeding in different areas and hence exploiting different stocks of fish. Another possibility is that they could utilize different age classes of fish. Puffins have smaller beaks than Common Guillemots, so it would not have been surprising if they had taken smaller fish. A third way they might reduce competition would be to exploit the fish at different times, either of day or through the season.

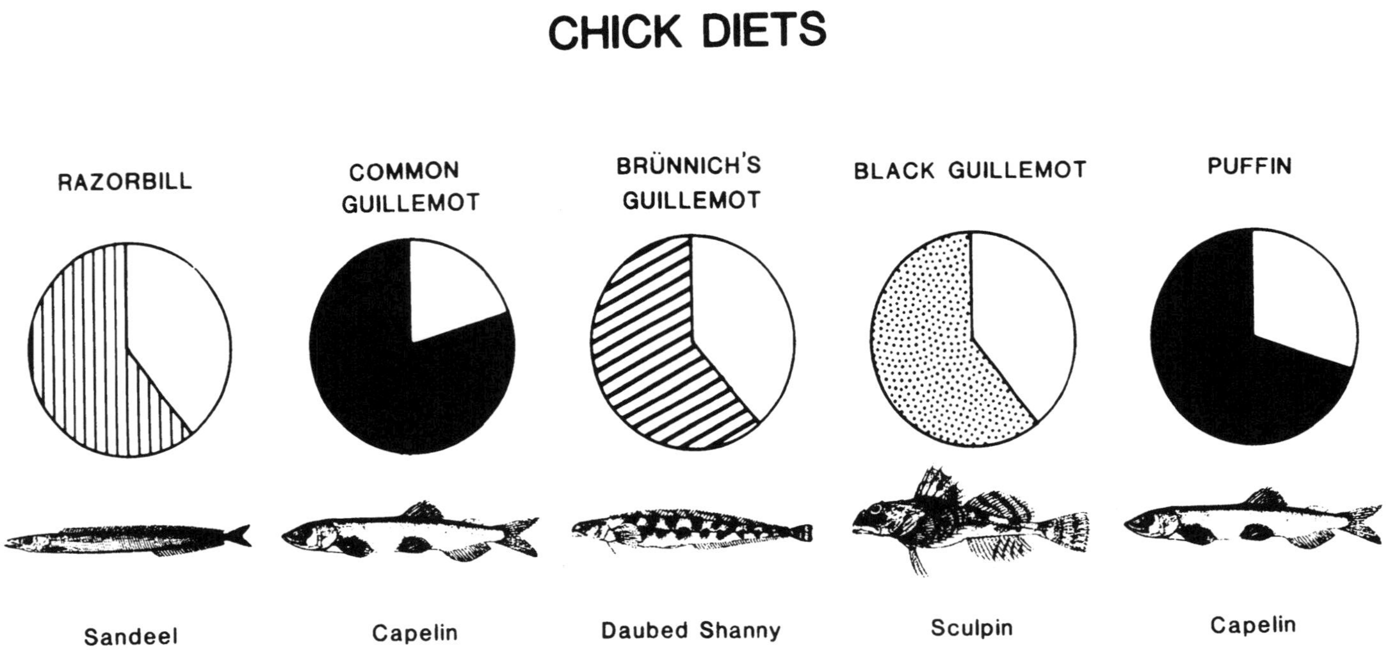

Fig. 19 *The diet of the chicks of the five auks breeding at the Gannet Clusters. Each species fed its chick predominantly on one species of fish (represented by the shaded portion of the pie-chart). Only the Common Guillemot and Puffin show any similarity in the fish species fed to their respective chicks.*

The boat-based surveys of Mike Bradstreet in 1981 showed that Puffins and Common Guillemots used the same general area for feeding, so we were able to exclude that possibility. Also, the average size of capelin that we obtained from the two species of bird about to feed their chicks was identical, so that too could be eliminated. This left 'time' as a segregating mechanism.

I then realized that in each year of our study, Puffins had reared their chicks later than any of the other species. This was partly because they started to lay their eggs later in the spring, but also because the duration of their chick-rearing period (38 days) is considerably longer than that of the two guillemots (21 days) and the Razorbill (17 days) (see Table 4.1).

By plotting the cumulative numbers of each species hatching, and subsequently fledging (Fig.20) it was clear that the chick-rearing period of the Puffin population at the Gannets overlapped hardly at all with that of the Common Guillemot, or with the two other large auks. We found too few Black Guillemot nests to include this species in our comparison, but it was not important in the present context because Puffins and Black Guillemots showed so little overlap in what they fed their chicks anyway (Fig. 19).

Rearing their chicks at different times of year (Fig. 20) may be the way that Puffins and Common Guillemots minimize competition between each other. But this was a rather weak way of testing the idea of ecological segregation. There were several alternative explanations for the difference in chick-rearing periods. For example, we knew that early in the season Puffins' burrows were often plugged with ice. They could not start to lay their eggs until the ice had melted, so this seems reason enough for their relatively late breeding. However, there were two ways I could discover whether the difference in the timing was a consequence of some other factor or whether it was to avoid competition for food. The first method involved looking at the results we had obtained from the two species in slightly more detail. The separation between the chick-rearing periods of Common Guillemots and Puffins was not absolute—there was a short period when Puffin chicks were just hatching and Common Guillemot chicks were preparing to fledge. This overlap period was the key to sorting this out. By looking at what each species was feeding to its young during this overlap period, and comparing it with the situation both before (i.e. when there were just Common Guillemot chicks at the colony) and after (when there were only Puffin chicks at the colony), it might be possible to test whether the difference in the chick-rearing periods was a way that Puffins and Common Guillemots avoided competition for food.

I predicted that if the Puffin's late breeding was simply the consequence of ice in their burrows, then we would have no reason to expect a difference in the diets of Puffin and Common Guillemot chick between the overlap period and the period before or afterwards.

Often when one is doing fieldwork there simply is not time to plot out the results as they come in. That was exactly our situation during the chick-rearing period since we spent virtually all our time doing feeding watches, weighing chicks to see how they grew, obtaining fish specimens and so on. At that stage all I knew was that both Puffins and Common Guillemots fed their chicks mainly on capelin and on capelin of a similar size. I also knew that the

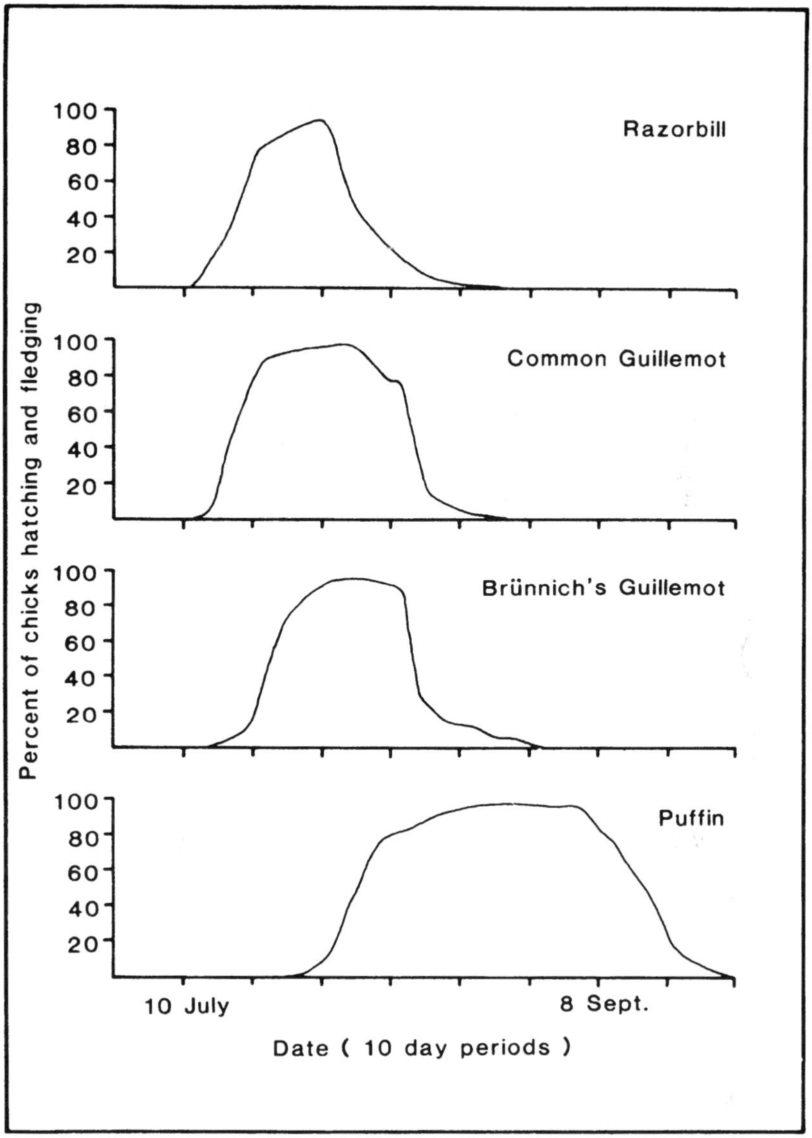

Fig. 20 *The timing of the chick-rearing periods of four auk species at the Gannet Clusters. The curves show the period when the chicks of each species are at the colony. The figure shows the cumulative number of chicks hatching (in July) and their subsequent departure at the end of the breeding season as they fledged (in August or September). Note that the chick-rearing periods of the Razorbill and the two guillemots are all similar, but that Puffins start to hatch much later, so much so that the majority of the former species have fledged relatively early in the Puffin's chick-rearing period (see text).*

rest of the Puffin chick's diet was composed of larval fish, and the remainder of the Common Guillemot chick's diet comprised a variety of other species: the stout eelblenny, daubed shanny, atlantic cod, arctic cod, fourline snake blenny, fish doctor, sandeel, eelpout and sculpins.

It was not until I tried to test my idea that I discovered, that for both the Puffin and the Common Guillemot, that these other fish, which in terms of their calorific value were generally of poorer quality, were utilized predominantly during the overlap period. This suggested that when both Puffins and Common Guillemots were foraging for their offspring at the same time, the competition for capelin was intensified, so that each species switched to alternative prey during that time (Fig. 21). The fact that the same pattern occurred in both years (1982 and 1983), even when breeding was significantly later for both auks in 1982, adds support to this interpretation.

The second approach to testing the idea of ecological segregation between Common Guillemots and Puffins was to compare the situation at the Gannet

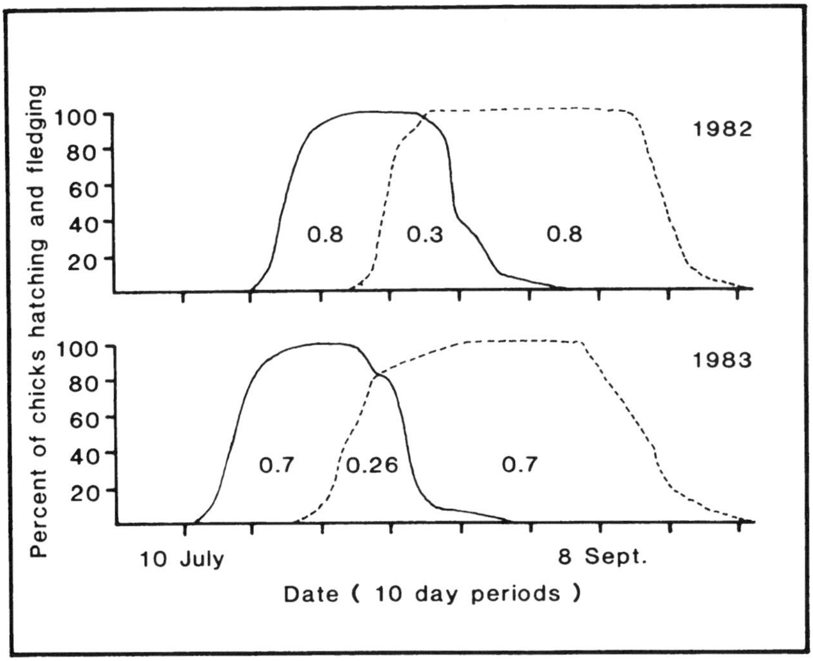

Fig. 21 *The overlap of the chick-rearing periods of the Common Guillemot (solid line) and Puffin (dashed line) at the Gannet Clusters (see Fig. 20), in relation to the similarity of their chick diets in 2 years. The bold numbers indicate the similarity of the diets (on a scale of 1 to 0, where 1 = identical, 0 = completely dissimilar – see text). Note that in both years the similarity of the diets was least when the chicks of both species were at the colony at the same time. This difference suggests that during this overlap period Common Guillemots and Puffins feed their young on different species to avoid competing with each other. Outside this period, i.e. before Puffin chicks have hatched or after Common Guillemot chicks have fledged.*

Clusters with that of other colonies. Common Guillemots and Puffins have been studied at several colonies in Britain (e.g. the Farne Islands: Pearson 1968; Skomer Island: Harris 1970; Foula: Furness 1983; Isle of May: M. P. Harris pers. comm.) and in all cases the situation differs markedly from that in Labrador. At the British colonies the chick-rearing periods of the Common Guillemot and Puffin overlap completely, the chicks are fed on either different fish species, or where the same fish are taken, the two auks exploit very different size classes. In other words, where the breeding seasons are similar the chick diets differ. In contrast, in Labrador, the chick diets are identical, but the breeding seasons differ.

Taken together, this information, along with the comparison between species at the Gannet Clusters (see Fig. 19), provides clear circumstantial evidence that Common Guillemots and Puffins avoid competing with each other in one way or another. Of course, making inferences about ecological processes from purely observational studies that rely entirely on natural variation in animals or environmental conditions can be dangerous. In the absence of experiments it is often difficult to disentangle the various factors that might cause a particular effect. In theory I could have tested the idea that Puffins breed late to avoid competition with Common Guillemots by removing all the Common Guillemots to see if Puffins subsequently started to breed earlier. Clearly such an experiment would be impractical (and somewhat unethical).

Community ecologists have been divided over the value of observational versus experimental studies. Some believe that it is possible to interpret observational studies that rely on so-called 'natural experiments', whereas others maintain that the only way to study communities is through a genuine experimental approach. Clearly though there are certain systems, such as seabird communites, where it is downright impossible to perform an experiment. In such cases one has two alternatives: either to do as I did with our Puffin–Common Guillemot problem and make the best use of the available information and recognize the limitations of the results, or simply not to bother looking at this type of problem at all. I sympathize with both views, acknowledging that experiments are the most effective way of testing biological ideas, but at the same time I cannot help feeling that it would be silly not to study a system simply because one cannot manipulate it in order to conduct an experiment. If we adopted that approach all ecologists would be studying single-celled animals in test-tubes.

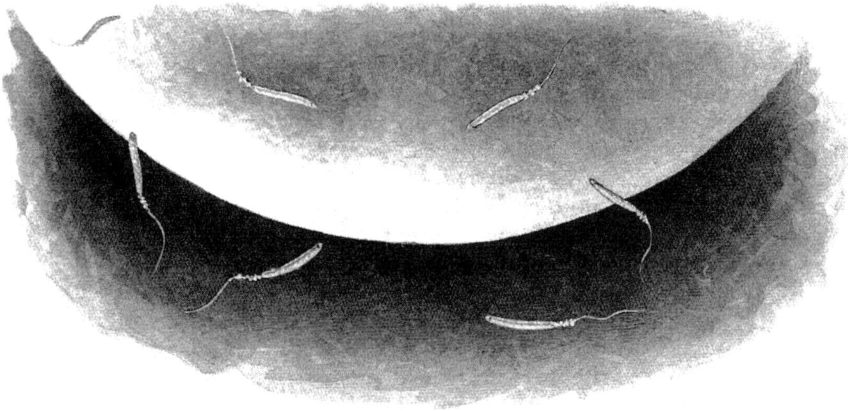

CHAPTER 8

The Fertile Sea

I have frequently observed promiscuous copulation at both common and thick-billed murre colonies. This occurs especially when the female is left temporarily unattended by her mate.

Tuck (1961)

Since the earliest records it has been apparent that spring and summer conditions on 'The Labrador' vary enormously from year to year. The amount of ice pouring south on the Labrador Current from Davis and Hudson Straits and the amount of landfast ice affects all those trying to wring a living from the sea, birds and man alike. In our second spring (in 1982) the ice conditions were bad, although nothing like as severe as those in 1978 (see Chapter 2). This was fortuitous in many respects, including the fact that it gave us a feel for how marine birds coped with different conditions. In that second spring most of the team, including Eric who remained with the project for all 3 years, arrived at the Clusters ahead of me. My teaching commitments in Sheffield kept me away until early June. However, when I did finally arrive, contrary to my expectations none of the birds had laid so I was able to witness some of the pre-laying events that had so far eluded me in Labrador.

The extensive sea-ice ensured that it was a cold spring and we were glad of our double thickness, down sleeping bags at night. Sitting for long periods in the hide was potentially unpleasant, but by donning several layers of pull-overs, hat, gloves and long johns and using improvised hot water bottles (the plastic and foil container in a wine box), we made ourselves comfortable.

The study plots were almost completely covered by snow and ice, which in some places was almost a metre thick. Guillemots of both species have amazing spatial awareness and although they could not see their sites all the birds sat on the ice directly above their correct positions. We were able to tell this from the general configuration of the birds on the plot, but confirmed it by the position of bridled individuals. As time went on some birds gradually sank into small pits in the ice, created by the warmth of their body. The ice on the study plot, coloured by the birds' droppings, was bright pink, with occasional streaks of yellow. On closer inspection one could see the fragments of crustacean exoskeletons that produced this lurid pink in their faeces. Interestingly, the colour of the Common Guillemots' droppings changed through the season, reflecting the changes in their diet. Early in the season it was crustacean pink and gamboge yellow, presumably from the fats in gravid capelin; later on, and immediately after egg-laying, female Common Guillemots often produced deep green droppings. During the chick-rearing period the droppings were predominantly white, presumably the calcium from ungravid and unfatty fish.

Despite the cold, the days before the birds started to lay eggs were both interesting and extraordinarily peaceful. The covering of sea-ice resulted in spectacular sunsets, and the ice also depressed any sea swell, so commuting between the islands was straightforward. Sitting in my hide and gaining

A late spring. Two Common Guillemots incubate on snow: their sites lies on the rocks beneath the snow.

familiarity with the study plot I started to appreciate the increasing potential for using the bridled birds and their site fidelity to recognize individuals. In behavioural ecology, ideas are what count and asking the right question is probably the hardest part of any research project. Back in Sheffield, my busy teaching schedule rarely gave me enough time to think, but sitting in a hide on a cliff top for hours at a stretch was ideal. As ideas came to me I noted them down in the back of my field note book. I would then worry them to death, thinking up alternative explanations, and trying to pick holes in them. Not surprisingly most ideas succumbed, but occasionally one would survive, and I would know I had the germ of something worth pursuing. Once you had a idea it was usually relatively easy to collect the necessary information to test it.

There was one idea I had had several years previously, but had not been able to test because I did not have the right set-up in the field. That spring I began to realize that things here in Labrador might be just right for exploring an idea concerning the promiscuous behaviour of Common Guillemots. If you sit and watch a group of Common Guillemots during the early part of the season, the chances are you will see birds of both sexes copulate with several different individuals in the space of a few hours or even minutes.

What is going on? This is a monogamous species, with long-term pair bonds, but here are males, and females too, fooling around with their next door neighbours. Other field biologists had witnessed similar 'promiscuous' behaviour in different species, but had usually written it off as aberrant behaviour: the male had a hormone imbalance, or was sexually dissatisfied with his partner.

* * *

A late spring: a pair of Common Guillemots copulate on the snow.

Instead of writing off Common Guillemot promiscuity as an aberration, the behavioural ecology approach encouraged me to ask 'how could this behaviour benefit the individuals performing these extra-pair matings?' I had asked myself this question many times during my PhD studies on Skomer as I watched Common Guillemots copulating with their neighbour's partner but then go on to rear a chick with their own long- standing mate. However, I had been perpetually thwarted by my inability to even begin to answer it because I needed birds I could recognize individually. Although I caught and marked many Common Guillemots on Skomer, I never had enough in any one colony to make it worthwhile trying to study their promiscuity.

As I watched from my hide on GC4 during that cold spring I began to realize that the perfect opportunity for studying extra-pair copulations was unravelling before my eyes. I said earlier that having a good idea is the focus of any research programme, but in fact a whole succession of ideas is necessary: a central one and then lots of others stemming from it. My main idea was to try and determine whether the extra-pair copulations were adaptive for males. The prediction arising from this is that if they are, then extra-pair copulations should occur only, or most often, when the female could be fertilized. Subsidiary questions included things like which sex initiates an extra-pair copulation? What proportion of extra-pair copulations are successful, compared with, say, copulations taking place within pairs? How did males respond to seeing their own partner involved in an extra-pair copulation and hence the risk of being cuckolded?

The main Common Guillemot study plot on GC4 where most behavioural observations were made. This picture was taken on a day when most of the birds were incubating and there were few off-duty birds present.

In the evening after supper, I would sit down with a block of graph paper and a pocket computer to plot out the information we had collected so far and test a few ideas. At the end of each day I would add in the new data, anxious to see if my predictions were holding up. I knew from the outset that it would be impossible to collect all the data I would require in that particular season, because the Common Guillemots started to lay their eggs within a few days of my seriously trying to collect the right information. Nonetheless, I knew that I could recognize enough birds to make such a study feasible.

For the rest of that summer I could not leave the idea alone, and the back of my note books filled up with plans for the following season. My main worry was a logistic one, of being able to get out to the Clusters sufficiently early in the season to make sure I collected all the information necessary. There were two logistic problems: first, I would normally still be teaching and marking exams in May, and second, access to the Gannet Clusters at that time of year would be unpredictable without a helicopter. Luckily I had a sympathetic head of department, who gave me his blessing and allowed me not only to leave my various duties behind in May, but also gave me the following October–December term as study leave. David Nettleship agreed that the project could stand the cost of a helicopter flight from Goose Bay to ensure that I got to the Gannets at the right time.

* * *

Throughout the winter I re-thought the ideas through, reading all the papers I could find in the scientific literature on this topic. Most of the studies I read about had been conducted on insects, and followed on from Geoff Parker's pioneering studies in the early 1970s. Geoff coined the term 'sperm competition', which is the process that occurs when two or more males inseminate a single female during one reproductive cycle. The sperm from these different males compete to fertilize the eggs of that female. Obviously, the sperm themselves are not conscious of competing, but the male whose sperm 'wins', that is, fertilizes the female's eggs, will be rewarded by having his genes passed onto the next generation. In contrast, the males that 'lose' in this competition will not. It is possible that the winning male in this situation has faster swimming sperm, but his success may have nothing to do with his sperm. It is equally plausible that the behaviour of the winning male differed in some way from that of the other birds. For example, he might have timed his insemination more precisely—just at the time when the female's eggs were ready to be fertilized. Thus the term sperm competition encompasses events occurring within the female's reproductive tract and behavioural events taking place up to and including insemination.

Even though behavioural ecologists had known about sperm competition since the early 1970s, there were very few studies of this topic in birds, even by the early 1980s. I am always surprised at how long it took ornithologists to take up the challenge. As a result, I read the literature on sperm competition in insects. This is not as perverse as it might at first seem because one of the most satisfying aspects of behavioural ecology is that the theory is not restricted by taxonomic groups. If female insects tend to copulate with several different males, then the ways in which males try to protect their paternity may also be applicable to birds.

* * *

During that winter I read extensively, and tried to put my thoughts into a broad perspective, fitting them into the framework of theoretical ideas. With the benefit of the previous season's data I had a pretty good idea of the type of information it was possible to collect, and also the likely outcome of my observations. This enabled me to design a standard chart onto which I could record my observations. It would then be possible to record data rapidly in a simple code-like form, which I knew would be essential since behavioural events often happened very rapidly. There is usually insufficient time to write lengthy descriptions of behavioural interactions as much as one might want to. An alternative approach I considered was using a tape recorder, and while this would have ensured that I missed nothing, I dreaded the idea of transcribing notes from tapes. I had done this previously for other behaviour studies and it doubles the amount of time necessary to obtain the required information. In the event I took a tape recorder with me, just in case, but fortunately I rarely had to use it.

Because it was possible to anticipate the type of results I would obtain I was more or less able to plan the scientific paper I would eventually write. This is one of the major advantages of working in an area where there is a good theoretical background, and where you have identified the questions you want to answer. By mentally writing the paper before even starting to collect data, it is possible to pinpoint possible shortcomings, or gaps that would make the account less complete. The trick is to write the paper in your brain, and then pretend you were someone else reading it. The 'someone else' always had to be one's most critical scientific colleague, and this could be either a real person or simply an imaginary one. In this way, over days and weeks I refined and revised my ideas until I reached the point where I was not making any further improvements. At this stage I knew I was ready to start.

* * *

Mentally prepared and full of anticipation I set out for Labrador the following May. The weather in Goose Bay was perfect, and it was a real treat to take a helicopter direct from there to the Gannet Clusters with none of the usual hanging around. In one sense it was uncanny: one day I was at home in Sheffield and the next I was flying over a forbidding sea full of huge sheets of ice towards the Gannet Clusters.

I had anxiously asked in Goose Bay what the ice conditions on the coast were like, since this could have considerable bearing on the project. Reports were mixed, some people told us that the ice had all gone, others said it was 'real bad'. I decided to wait and see for myself, and it was with a sense of relief that I saw that while there was extensive landfast ice along the coast, the water

Flying over the Gannet Clusters: large numbers of Common Guillemots near GC1.

around the Clusters was almost entirely ice-free. Most importantly, there were large numbers of Common Guillemots sitting on the water in between the islands. Everything suggested that this was an average ice year and that the birds would breed on time.

A rapid check revealed that the cabins on GC2 and GC4 had survived the winter, and that the hides were all in good shape. We used the helicopter to move equipment from GC2 to GC4 and then sorted things out in the GC4 cabin, ensuring that the propane stove and the radio, our link with Cartwright, worked satisfactorily.

This was the first spring I had spent at a large Arctic seabird colony, where I had arrived at about the same time as the birds. I was fairly certain the birds had only recently returned to the Gannet Clusters because there were virtually none of their colourful droppings on the breeding areas, suggesting that few individuals had so far ventured onto land that season. The satellite photos of the Labrador coast that I subsequently examined confirmed that a few days prior to our arrival the ice in the whole of the Sandwich Bay area had been continuous.

The two guillemot species, Razorbills and Puffins remained on the open water between patches of ice. They came and went, disappearing to the south, presumably to feed, then returning in enormous, wheeling flocks to settle on the water. Each flock consisted of tens of thousands of birds, and one could almost feel their excitement about returning to their breeding areas. Earlier, I disparagingly described Brünnich's Guillemots as phlegmatic and rather dull compared with Common Guillemots, but this was one part of their pre-laying behaviour where they eclipsed their more extrovert cousins. As the days progressed, the flocks of birds spent more and more time near the islands and got ever closer to the cliffs. Usually both species of guillemots occurred together in the same flocks, but on some days Brünnich's Guillemots were there alone or in separate flocks.

Several hundred Brünnich's Guillemots sat on the water, mostly silent, and occasionally preening themselves. As I watched, the atmosphere in the flock began to change, and an increasing number of birds started to dive together and then stand up in the water and wing-flap. Over a period of about 10 minutes, more and more birds started to do this until the sight and noise of all this movement and splashing reached a climax. At this point large numbers of birds began to skitter across the water surface to take off, and almost as though in response to an invisible signal, the entire flock was airborne, and flying close in towards the breeding cliffs. They flew along the cliffs in large groups and the sound of their rushing wings was audible as they flew past. The birds were apprehensive and appeared to be uncertain whether to land or not. They flew round in a huge circle, flying close to the ledges and then turning away at the last moment. Then, as they came past again, almost synchronously they began to alight on their breeding ledges. This was the first landfall this season and as partners were reunited there was an explosion of noisy, growling greeting calls and a frenzy of copulation activity. You could almost feel their ecstasy at finding their partner from previous years. Like Common Guillemots these are long-lived birds and many retain the same partner for their entire breeding life. It is hardly surprising that they were pleased to see each other.

A number of long term studies of seabirds have shown that breeding success of individuals increases with experience, but also with the duration of the pair-bond. Partners get to know each other's foibles, and develop a coordinated rhythm of incubation and chick-feeding. If two birds have developed a good system and have been successful in rearing offspring, it is very much in both their interests to re-pair with the same bird the following year. One reason why long-lived seabirds re-use the same breeding site year after year is simply to provide a place where partners can meet.

The land-coming of the Common Guillemots was rather less spectacular than that of their Arctic cousins. This was because the Common Guillemot's breeding area was low down by the sea, most birds arrived, not by air, but from the sea. They swam close inshore, and then hopped up over the ice and onto their rocky sites.

* * *

I decided to watch the Common Guillemots for two 2-hour periods each day, one in the morning and one in the afternoon. This would allow me to collect sufficient information, but also give enough time to summarize each day's data (and do the various other things that were necessary for the other parts of the Gannet Clusters' auk project). Any field study requires a sustained effort, particularly if you are keen to obtain tidy data. Having decided on 4 hours of observation each day I was committed to this schedule for several weeks until all the birds in the study area had laid. Although a 2-hour period of observation sounds rather short, it involved intense concentration since you could not afford to let your mind wander in case you missed something crucial. Moreover, the action among the birds on the study plot was sufficiently intense that I was able to collect plenty of information during this time.

In other areas where I had watched Common Guillemots during their pre-laying period, such as on Skomer Island, the birds showed a very regular pattern of presence and absence at the colony. Usually, the birds were present for 3 days then away, foraging, for 3 or 4 days. At the Gannet Islands this did not occur, possibly because food was available closer to the colony, and birds were present at their breeding sites every day. However, as at other colonies, the birds were initially extremely nervous and the slightest disturbance would send them panicking from the breeding areas to the safety of the sea. One morning soon after I had started my observations a Peregrine drifted over the island, flying at eye-level a few metres away from my hide window. The Common Guillemots abandoned the colony with undignified haste, leaving me with mixed feelings. It was good to see the Peregrine so closely, but I realized that it would take an hour or two before the birds would return. Moreover, I knew I must remain in the hide during that time in order not create any further disturbance.

Once on the sea a few hundred metres offshore the birds relaxed, presumably feeling safer there because they could dive to safety from a predator like a falcon. They resumed their preening and bathing, and almost imperceptibly started to drift back towards the colony. After what seemed like a long wait to me, the first bird would hop ashore and tentatively make its way towards its site. The others soon followed, until the entire colony was back where they started and I was able to start collecting information once more. Sometimes, however, the Peregrine would return and the whole process would start again.

Initially all birds left the colony at dusk each evening and roosted out at sea some distance from the colony. I assumed that they fed either last thing at night or in the early morning before returning soon after first light. After a few days however, the birds started to remain at the colony all night, something I had never seen among Common Guillemots on Skomer at this stage of the breeding cycle.

*   *   *

In the previous spring I had realized from my observations of bridled birds that the sex ratio at the colony just before egg-laying started was far from even, and I was now anxious to monitor this closely. Knowing about the sex ratio was important for several reasons, but particularly because I wanted to see how it affected the likelihood of extra-pair copulations occurring. I predicted that when there were lots of randy males around and only a few potentially fertilizable females, the pressure on females would be greater than at other times. In fact, on the first day in the hide at the beginning of the season I found that at most sites both pair members were present and there was therefore an equal sex ratio at the colony. On their first return to the breeding areas in the spring it is presumably in the interests of both partners to let the other one know that they are alive and kicking, to reduce the risk of either one trying to establish a bond with another bird.

This was beautifully illustrated by one particularly distinctive pair of birds on my study plot. The male of this pair at site No. 250 had been present since the start of the study, and although he was unmarked he was instantly recognizable because he retained his winter facial plumage throughout the summer. For some reason he simply did not moult into the uniform brown head of typical summer plumage. This year however, I looked for him in vain. His partner turned up at the site and she sat there alone. I was saddened by male 250's disappearance but assumed that he had been one of the unfortunate few percent that failed to survive the winter. After about a week of the female being on her own a new male turned up—they preened each other and copulated, and generally started to behave as a normal pair.

Then one morning about 10 days later, I looked out of the side window of my hide towards the birds on the sea just offshore, and saw a winter-plumaged Common Guillemot swimming purposefully towards the colony. I immediately thought that this must be male 250, and as he got closer to the colony I became more certain. I watched with anxious anticipation. His partner and her new lover were both at the site (his old site), as male 250 hopped out of the water and started to make his way through the colony towards them. Even when he was 5 m away his female recognized him and she started to call to him. As they met their greeting display was one of the longest and most intense I have ever witnessed. Both birds uttered their long, guttural '*aaargh*' as they clashed their open beaks together. The new male stood to one side, no doubt watching in disbelief. Male 250 suddenly appeared to notice him, put two and two together, and stepped over and gave him a couple of hefty blows with his beak and drove him off. He never came back. Meanwhile, pair 250, reunited, settled down for a long session of mutual preening, as though to cement their re-formed pair bond.

* * *

As time went on the sex ratio became increasingly biased towards males with fewer and fewer females at the colony (Fig. 22). The females were away at sea foraging to form their single, relatively large egg. A Common Guillemot's egg weighs about 113 g, which is 11–12% of the female's body weight, slightly less than a human neonate, which weighs about 16% of its mother's weight. The period when females are forming their eggs is one we know little about: the females are out of sight at sea for much of the time, returning to the colony occasionally simply to copulate with their partner. Most of their time must be spent foraging to allow them to accumulate the nutrients to form their egg.

In fact this period of feeding is used to form the true egg, or ovum. This is the yolk plus the germ cell, which will subsequently be fertilized. Only after the ovum is fully formed is the albumen laid down around it and finally the shell put on the egg. The complete ovum weighs approximately 30 g and is the size of a ping-pong ball. By examining the yolk of newly laid eggs we can

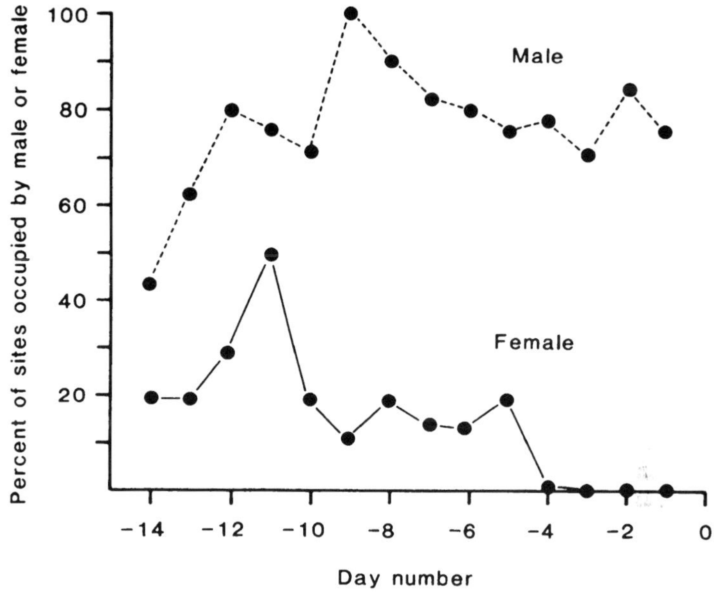

Fig. 22 *Differences between male and female Common Guillemots in the time they spent at the colony in the 14 days prior to egg-laying. The horizontal scale shows days before egg laying (where day 0 = the day of laying). The vertical axis shows what percentage of the 51 individual sites under observation was attended by each. It is clear that males were present at their site more often than females, presumably because females were away foraging in order to form their egg (see text).*

trace the events leading up to egg-laying, and discover how long the female took to form her ovum (Fig. 23).

While the female Common Guillemot is at sea she feeds on a mixture of fish and crustaceans. The nutrients in that food are synthesized and laid down in her ovary as yolk. During the day when the female is busy foraging and actively producing yolk from the food she ingests, the yolk laid down is dark and dense. In contrast, the yolk laid down overnight, when the female is not foraging, is relatively light in colour. The difference between the yolk formed during the day and the night can be detected by placing a slice of the entire, fully formed yolk from a freshly laid egg in a solution of potassium dichromate. This enhances the day–night difference and the number of pairs of dark–light rings can be counted to determine exactly how long it took to form the yolk (Fig. 23). In Common Guillemots this is about 14 days on average.

The female's ovary contains numerous follicles, all of which have the potential to develop into a complete egg. Since guillemots, and most other auks, lay only a single egg we might expect only one follicle to develop each year. I dissected a few birds that I had found dead just before egg-laying and found this to be generally true, but some had also started to develop a second

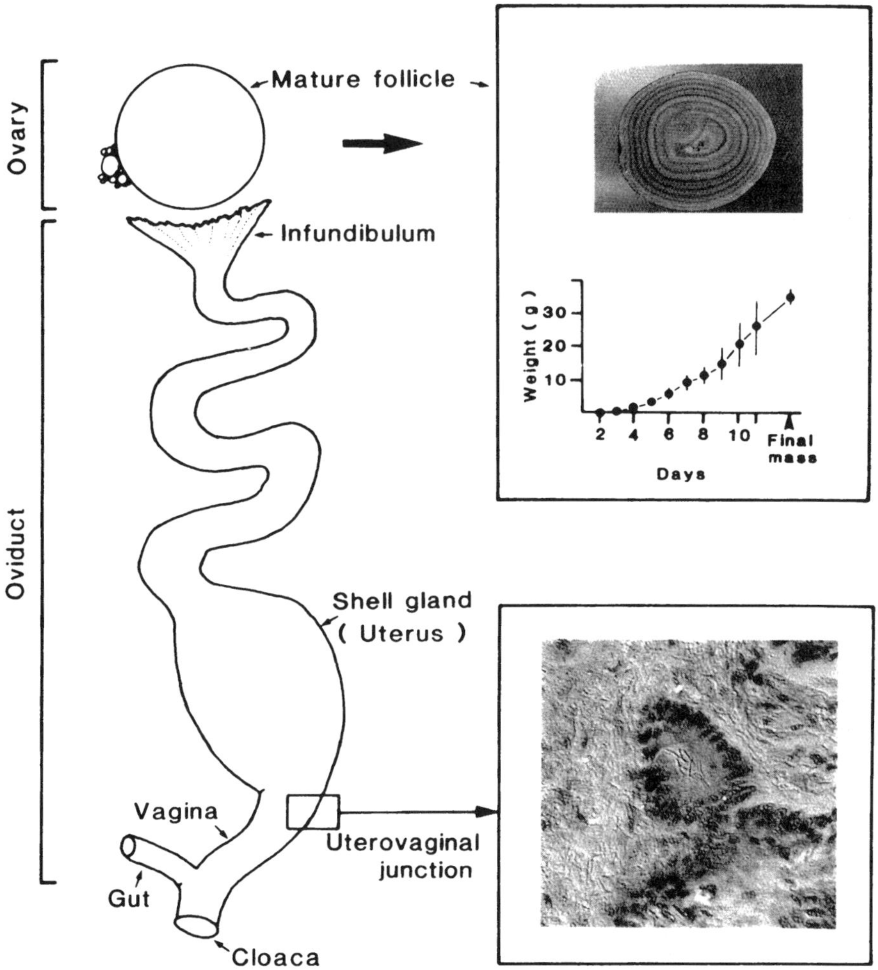

Fig. 23 *The reproductive tract of a female Common Guillemot showing the major regions. The upper insert shows the alternating light and dark bands indicating diurnal patterns of yolk deposition in a fully developed yolk follicle. The graph shows the pattern of 'growth' of the yolk during its formation (from Birkhead and del Nevo 1987). The lower insert shows a cross section of one the many sperm storage tubules (actual diameter is one-fifth of a millimetre) located in the utero-vaginal region: the dark bar-like structures are the heads of the spermatozoa lying within the tubule.*

follicle. This was not because they were going to lay a clutch of two eggs, but, presumably as an insurance in case the first egg was lost. The fact that another follicle had already commenced its development would minimize the delay in laying a replacement egg.

Growing the ovum is only part of the process of egg formation. After fertilization has taken place, the albumen and shell are laid down around the yolky ovum. The sequence of events is carefully regulated, and has important implications for understanding the copulation behaviour of the birds.

Once the ovum is complete it is retained within a small bag of tissue, the follicle, and it is probably held here for about 4 days (Birkhead and del Nevo 1987). No-one is quite certain why there is this 4 day delay, but it may be to allow the female to accumulate the nutrients for the other part of the egg, the albumen. After the allotted time, the bag splits and the ovum is released into the top of the female's oviduct. It is here that sperm are waiting, and they soon swarm all over the egg's surface until one lucky one finds the germ cell and fertilization takes place. Within 30 minutes of being shed from the ovary the first layers of albumen are laid down around the yolk. If a sperm has not fertilized the ovum by this point, the ovum will remain unfertilized because sperm are incapable of penetrating the albumen. What this means is that the time period when fertilization can take place is remarkably short, probably no more than an hour.

The fertilized ovum then continues on its journey down the oviduct, first being surrounded by albumen, and then, in the shell gland, having two tough membranes, one on top of the other laid down around the albumen. Once these are in place the shell is gradually built up. The beautiful blue, green or white background colours of the guillemots' egg are produced in the main body of the shell. Finally the intricate dark patterning is laid down on the outermost part of the shell. The egg's colours are produced from the break-down products of body fluids: the blue and green ground colours are derived from bile while the red and brown marking colours are synthesized from blood.

In contrast to Common Guillemots and most other seabirds, female songbirds, such as the European Starling, produce clutches of four or five eggs, laying one egg each day. The four or five follicles develop simultaneously, but one day out of phase with each other. Each ovum is shed at 24 hour intervals, and each fully formed egg is also laid 24 hours apart. We thus know that the business of laying down the albumen and the shell takes one full day. In guillemots and other species that lay a single egg, we have no idea of how long these processes take. It is certainly not less than 24 hours and may well be 2 days or more.

Like many other seabirds, Common Guillemots copulate over a period of several weeks prior to egg-laying. If the interval between the ovum being shed from its follicle and fertilization is so short (above), why do birds start to copulate weeks in advance of this? There are several interrelated answers to this question. At one level the answer is straightforward: because the time period when the ovum can be fertilized is so incredibly short, birds need to copulate in advance so that the female can store the sperm and have it ready for fertilization when ovulation occurs. Until fairly recently it was not generally known that birds had the ability to store sperm. Indeed, one otherwise outstanding ornithological textbook published in 1990 states explicitly that 'Female birds have no specialized sperm storage areas'. In fact the first record of female birds storing sperm was made over one 100 years prior to this (Tauber 1875).

All the species of birds examined so far have specialized sperm storage tubules located in the wall of the reproductive tract at the junction of the vagina and uterus (or shell gland). I was keen to discover whether, as I strongly suspected, Common Guillemots also had these tubules. A colleague who was working in Alaska, where the local people shoot large numbers of seabirds to eat, kindly responded to my request to send me a reproductive tract of a female Common Guillemot. The material was not in great condition, probably because it had sat around for several hours before being placed in the zoologists' standard, smelly preservative, formalin. Nonetheless, I was able cut thin sections of this tissue and examine them under the microscope, revealing that Common Guillemots did indeed possess sperm storage tubules (Fig. 23). The most likely sequence of events is that the Common Guillemots in Labrador start to copulate from the day they first meet at the colony, some 3 weeks before the onset of egg-laying. Sperm from these copulations travel up the female tract and some of them are stored in the tubules. For reasons not yet understood, females allow only a small subset of the inseminated sperm into their tubules, the rest being ejected from the cloaca. With successive copulations sperm accumulate in the tubules and then at the right moment are released so that they travel to the farthest end of the oviduct, to a funnel-shaped region referred to as the infundibulum. The sperm accumulate here, ready to fertilize the ovum when it is shed from its follicle in the ovary. It is not known just when the right moment is for sperm to be released from their storage tubules, nor is it known what it is that triggers their release. Most biologists who are interested in the reproductive physiology of birds study chickens or turkeys, but even in such well-studied species these aspects of their inner workings are still a mystery.

* * *

To be sure that they have sperm ready and waiting to fertilize their egg at precisely the right moment, the female has to return to the colony to copulate. She has no other option since this is where her partner is waiting. Why he waits at the colony for days or weeks on end just for a few copulations with his female will shortly become apparent.

Unfortunately for the females, returning to the colony and their mates is fraught with problems. Unless the female can alight directly beside her partner she has to run the gauntlet of numerous males, many of whom think they can coerce her into mating with them. Even if a female is skilful enough to alight at her own site, neighbouring males may still attempt to copulate with her forcibly. The male partner defends his female vigorously and he aggressively attacks any male that attempts to mount his female. Occasionally however, females return to the colony to find their partner absent. Under these circumstances they are particularly vulnerable to the forced extra-pair copulation attempts by other males.

In about one-third of extra-pair copulation attempts observed, more than one male tried to mount the female, and in 7% of cases between four and 10 males were involved. When this occurred the female was often pinned to the ground by the males, all of whom attempted to copulate with her. In these circumstances the male partner was totally incapable of defending his female. These melées looked unpleasant and potentially dangerous for females, but the males' over-zealous behaviour often made me laugh out loud. Males were so keen to secure extra-pair copulations that they would attempt to copulate with whatever part of the female's body was exposed, usually the head! On the other hand there were cases where the female was pinned down with her cloaca exposed and males then appeared to be successful in forcibly inseminating her.

Males remained at the colony throughout the pre-laying period and by adopting this pattern of colony attendance males were as sure as they could be of being there when their female decided to come back to copulate. The other big advantage, however, was that when their own female was not there (which was most of the time) they were free to attempt extra-pair copulations with the females of other males.

Males endeavoured to obtain extra-pair copulations in several different ways. About one-third of all attempts occurred through the males leaping onto an incoming female. Another third occurred apparently spontaneously: a male would suddenly approach a female that had been standing quietly at her site for some time, and try to mount her. In the remaining cases males tried to disrupt pair copulations. During a normal pair copulation the female falls forward onto her belly, giving a distinctive solicitation call as she does so. The male then steps onto her back and positions himself so that their two cloacas are juxtaposed. Extra-pair males seem to exploit the fact that the female appears to go into copulation mode, and they rush at the copulating pair, knock the pair

An extra-pair copulation attempt by a male Common Guillemot. The male is enthusiastic, but the female shows her reluctance by standing upright and trying to shrug the male off. A bystander (right) looks on with interest.

male (i.e. the rightful male) from the female's back and in a single deft movement substitute themselves. Sometimes when this occurred the female appeared not to notice (hard to believe) and continued to copulate. The extra-pair male's likelihood of success also depended on how hard he hit the pair male. If the pair male was hit hard and pushed over a cliff ledge by the extra-pair male then by the time he had recovered his senses and returned to the site the extra-pair copulation would be over. Anything less than this and the pair male would furiously attack the extra-pair male and disrupt his copulation.

Sometimes, instead of knocking the pair male off the female, an extra-pair male would attempt to prise the pair male off his female's back, by trying to squeeze himself between the two. This was an ineffectual strategy which usually ended up as a vicious fight between the two males.

Another tactic used by extra-pair males was to wait until a pair male was engaged in a fight with another bird, and then attempt to copulate with his female, while her partner was distracted and unable to defend her.

Interestingly, not all males attempted extra-pair copulations to the same

Extra-pair copulation attempts. If most birds had not been individually recognizable it would have difficult working out whom was doing what to whom in these melees.

extent. Of the 51 pairs I watched in detail, 17 males were never seen to be involved in extra-pair copulations, whereas eight males performed between four and 10. One explanation for this is that males differed in their physical strength and since most extra-pair copulations were fairly strenuous affairs, both in terms of subduing a female and having to tangle with her partner, perhaps only the toughest males even bothered to try.

There was no obvious difference in the relative attractiveness of different females as extra-pair targets: all females appeared to be equally vulnerable. In some ways this result was not surprising, since a male's strategy seemed to be 'shoot first—ask questions later'. So, a female would alight at the colony and a nearby male would instantly rush up to her and attempt to copulate with her. The males' behaviour followed a set pattern: on seeing the female arrive the male would start to 'crow', a harsh, yodelling call, and as he did so he would run towards the female, hook his neck around hers and then with scrabbling feet endeavour to mount. In 85% of the 237 attempted extra-pair copulation attempts I saw, the females resisted by either running away, attacking the male, or standing very upright so the male could not mount properly.

What was going on in the remaining 15% instances, when a female accepted a male's advances? First, females only seemed to go along with a male's extra-pair copulation if her partner was away. This may have been to avoid any possible reprisal by the male: for example, a male that thinks he might have been cuckolded could reduce the amount of effort he puts into rearing the subsequent offspring. If this occurred the female would suffer because she would have to increase her workload to compensate for his reduced effort. Although this might sound far-fetched it is exactly what happens in some other species (e.g. Davies 1992).

There are other possible reasons why a female might be prepared to accept the advances of an extra-pair male. First, it may be less trouble than fighting off a persistent male, and if her own partner is absent then there may appear to be nothing to lose by behaving in this way. On the other hand, if there was a chance that a female might be fertilized by an extra-pair copulation, then complying with a male just to avoid hassle would be a very risky thing to do, unless she knew that the extra-pair male was at least as good as her partner, in genetic terms.

Females may be happy to copulate with other males if they think their partner has died, as in the case of the female at site 250 described above. Indeed, copulation may be an important part of the pair formation process in Common Guillemots and a few other birds.

In some cases females may want to be inseminated by a particular male, but also to retain their partner. In a few instances I actually saw females initiate copulations with extra-pair males, by soliciting in front of them. This raises a tricky question: why bother to form a long-term pair-bond with one male, only to copulate sneakily with the next-door neighbour when your partner is not around? The answer seems to be that males (and females) differ in quality. As in humans, only a few females get the best males (and *vice versa*) as long-term partners, and the rest must make do with what they can get. When this occurs a female may attempt discreetly to have her egg fertilized by one of the good quality males, but continue the relationship with her partner as long as he is competent as a provider for her offspring.

Such an explanation may sound anthropomorphic. However, Common Guillemots or other animals do not think through their various options before deciding on how to behave. Instead, natural selection, and in this case sexual selection, operates in a purely mechanistic way. Individuals with one particular set of behaviour patterns end up with more copies of their genes in subsequent generations than individuals with a different set of behaviours.

In addition, these suggestions imply that to a large extent females control copulations. The reproductive behaviour of males is usually so much more obvious and lacking in subtlety compared with that of females that we end up with a biased viewpoint. The idea of studying extra-pair copulation in Common Guillemots was initially attractive to me because the behaviour of males was so obvious and easily recorded. However, the frequent and conspicuous attempts by males to force extra-pair copulations on females probably gives a misleading picture of what is really going on. Although I suspect that some males may have successfully inseminated females through forced extra-pair copulations, overall it is almost certainly females that run the show. It is largely they who determine whether any particular copulation is successful or not, and even if they are forcibly inseminated they may still have the ability to eject that male's sperm.

* * *

How successful were extra-pair copulations? This was the ultimate question I wanted to answer, but it was also the most difficult one. In the early 1980s when I was making these observations, it was not even possible to assign paternity in humans with any certainty. By using blood groups forensic scientists could say who was *not* the father of a particular child, but they could never say categorically who was. If it was not possible to assign paternity in humans, what were the chances in birds?

In fact the chances were rather less than they were for humans, and the technology necessary to do this was far from straightforward. I decided therefore to do the only thing possible in the circumstances and make some educated guesses. There are several possible ways in which sperm might compete to fertilize an egg, and the simplest is proportional representation. That is, the chances of the sperm from an extra-pair copulation fertilizing a female's egg is dependent on how many sperm from pair copulations it has to compete with. For example, if a female received one extra-pair copulation and one pair copulation, each male would have an even chance of fertilizing her single egg. On the other hand, if the female received one extra-pair copulation and 20 pair copulations, the pair male would be 20 times more likely to be the father of the resulting offspring. Of course, this assumes that similar numbers of sperm are transferred during pair and extra-pair copulations.

To evaluate this idea I needed to know how many times each female performed copulations with extra-pair males and how many times they copulated with their partner. Fortunately I had collected these data during my observations, and found that females copulated more often with their partners as the time of egg laying approached, from about one copulation every 5 hours at the beginning of the season to once every 2.5 h four days before the egg was laid. Interestingly, females were rarely present at the colony in the 4 days prior to the egg being laid (Fig. 22) and I wondered if this was the period of time between the egg being ovulated and laid.

It was also clear during my observations was that copulations occurred in waves. The presence of one pair of Common Guillemots copulating triggered off a succession of other pairs. Indeed, any kind of excitement in the colony, such as a fight or an extra-pair copulation attempt, had the same effect. I speculated about why this should occur, but could not come up with much other than that it might be a way of females synchronizing their egg laying (see Chapter 6).

So how many times do a pair of Common Guillemots copulate for their single egg? To answer this question in a sensible way we need one further piece of information, that is the length of time that females are fertile. Unlike the majority of mammals, female birds store sperm from copulations, and the length of time that sperm remain viable in the female reproductive tract determines their fertile period. Domestic Turkeys hold the record here and the average duration of sperm storage is 45 days, but there are a few cases of 112-day-old turkey sperm fertilizing eggs! In the majority of other birds the period of sperm storage is much shorter, just 6 days in the Ring-necked Dove, for example. Unfortunately we do not know how long sperm remain fertile in the female Common Guillemot. However, it is possible to use field observations to obtain a rough figure. There was one female on my study colony who visited the colony several times up until 11 days before she laid her egg, but then did not return again to the colony until they day she came back to lay. Her egg was fertile so we know that she had almost certainly stored sperm over this period. If the egg was fertilized the day before she laid (see above), then she had stored the sperm for 10 days. Ben Hatchwell (1988), who subsequently studied Common Guillemot extra-pair behaviour on Skomer, noted one female that went 16 days without copulating before laying a fertile egg.

To be on the safe side we assumed that any copulations, both pair- and extra-pair, occurring in the last 12 days before egg-laying would fall within the female's fertile period. During this time females copulated with their partner eight times and 0.4 times with other males on average. Overall, then about 5% of all copulations were extra-pair copulations.

The question we were left with is 'how do copulations translate into offspring?' Does an extra-pair copulation rate of 5% mean that on average 5% of all Common Guillemot offspring will be fathered by extra-pair copulations? There are at least two ways one can find out, neither of which were available when I conducted this study.

About 2 years after the completion of my study of extra-pair copulations in Common Guillemots and in the year in which it was published (Birkhead *et al.* 1985), a new method of assigning paternity was developed, by Alec Jeffreys at Leicester University, in England. Jeffreys was studying the human myoglobin gene, and as is so often the case in science, he discovered more or less by accident that sections of the DNA are virtually unique to each individual. Hence the name 'DNA fingerprints'. Jeffreys was a sufficiently good scientist to realise the potential of his discovery in terms of sorting out the problem of paternity, not only in humans, but also for behavioural ecology. However, it was Terry Burke, also at Leicester University, who really developed the technique for birds. He showed that in the common House Sparrow some offspring reared by a pair of birds were unrelated to the male (Burke and Bruford 1987).

It would be nice if I could say that once the technique of DNA fingerprinting had been developed, I was able to rush back to Labrador (or indeed any other Common Guillemot colony) and check how often males were successful at achieving extra-pair fertilizations. For better or worse, I have not (yet) done this. There are several things that have stopped me. First, fingerprinting is a sophisticated technique requiring considerable skill and a great deal of patience in the laboratory. In fact it requires far more ability and patience than I have, so that has been a major deterrent. Second, fingerprinting is expensive and extremely time-consuming, so it is not to be undertaken lightly.

Despite these deterrents, in the past few years there have been a number of studies of paternity in birds, most of them using DNA fingerprinting to assess the incidence of extra-pair paternity. My Danish colleague Anders Møller and I collected all the information together from these studies, and looked at the extent to which extra-pair paternity was related to the frequency of extra-pair copulation in the same species. To our surprise, and satisfaction, there was indeed a positive relationship between these two (Fig. 24). In other words, in

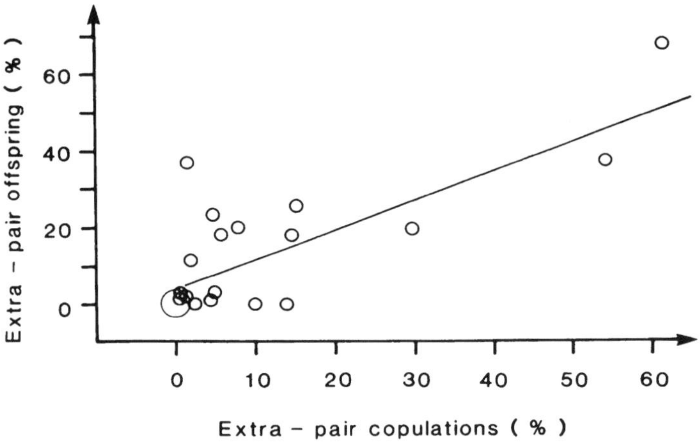

Fig. 24 *Relationship between the observed level of extra-pair copulation (the proportion of all copulations that are extra-pair) and the level of extra-pair paternity (the proportion of all offspring that are extra-pair) in 41 species of birds. (The large symbol indicates six species). From Birkhead and Møller (1992).*

species where females were seen to copulate regularly with males other than their usual partner the incidence of extra-pair paternity was high. At the other end of the scale, there were a number of species in which extra-pair copulation was seen only rarely, and no cases were recorded of any offspring being fathered by any but the presumed father.

This might not seem all that surprising, but we were relieved to find that it was true, for several reasons. First, it had been suggested by some behavioural ecologists that extra-pair copulations might differ from pair copulations in their potency. Males performing an extra-pair copulation might, for example transfer more sperm than they would during a copulation with their usual partner. Second, it had also been suggested that extra-pair copulations were more difficult to observe than the copulations between pair members. If this were true then it might be impossible to obtain an unbiased estimate of the proportion of all copulations that took place outside the pair-bond. However, as the results indicate, what you see as an observer of bird behaviour in terms of who mates with whom, and how often, provides a reasonable picture of the paternity situation.

What this means is that if we were to undertake a DNA fingerprinting study of the Labrador Common Guillemots we could expect about 5% of all offspring to be fathered by extra-pair copulations since these composed about 5% of all copulations.

If this was the only information we were after then it probably would not be worth launching into such a large and costly project. But the beauty of DNA fingerprinting is its precision in measuring the true degree of relatedness between individuals, and hence the true reproductive success of particular individuals. As an example, I said earlier that some males were especially active in attempting to force themselves on the females of other males. Through fingerprinting it would be possible to tell whether these males actually fathered more offspring as a result of their forced extra-pair copulations. So far, there is no specific evidence that forced extra-pair copulations result in fertilization. At the other extreme it would be possible to examine those cases where females actively solicited extra-pair copulations from males to see if they were more or less likely to result in offspring.

Because DNA fingerprinting can also allow us to establish the degree of genetic relatedness between individuals, it has the potential to answer those questions described in Chapter 6 about the Common Guillemot's social organization.

I have extolled the virtues of the DNA fingerprinting technique at some length, not simply because this is a new and trendy technique, but because the whole superstructure of behavioural ecology rests on being able to measure the true reproductive success of individuals. Prior to the advent of fingerprinting, ornithologists measured reproductive success of birds by simply counting how many chicks fledged from a particular pair's nest. Fingerprinting has shown us that this can be completely misleading, since in some cases an extra-pair male may have fertilized a female's entire clutch. As Fig: 24 shows, in some bird species such as the North American Indigo Bunting, as many as 40% of all offspring in a population may be fathered through extra-pair copulations!

* * *

You might be asking yourself at this stage,'what about humans?'. What indeed. In the past, paternity analyses using blood groups provided a rough indication of the incidence of extra-pair paternity in human populations, but as might be expected, fingerprinting has made those estimates much more precise. However, in contrast to behavioural ecologists who have gone out with the specific intention of measuring and announcing their estimates of extra-pair paternity in birds, the information from humans is much harder to get hold of. This is not surprising: imagine you could buy a do-it-yourself DNA fingerprinting kit from the local chemist or drug store. You make your purchase, go home and announce over supper that you are about to fingerprint your (putative) parents and (putative) brothers and sisters. The results are potentially explosive.

Many human diseases like haemophilia are known to be inherited, but there are others for which this is not yet known, so it may not be surprising to learn that there are several projects under way to investigate this. Earlier studies of this sort had to rely largely on people's statements about who their father was. Maternity is much less often called into doubt because females are obviously present at the birth of their offspring. In a research programme at a hospital in southern England, designed to determine the heritability of certain diseases, DNA fingerprints have been made routinely but secretly from patients' tissues. This enables the investigators to exclude instances where there is a mismatch between putative relatedness between individuals and the inheritance of the disease. Of course the information from the DNA fingerprints also tells us what proportion of offspring in this subset of people have been fathered by males other than the putative father. The surprising result is about 10%. Unfortunately (or fortunately, depending upon your perspective) results such as these have not been published, and therefore are not available for scientific scrutiny (see Macintyre and Sooman 1991).

Another unpublished study, conducted in a tenement block in northern England, apparently found that no fewer than 30% of children were fathered through extra-pair liaisons. These two examples provide an interesting parallel between birds and humans. Common Guillemots form long-term, monogamous pair-bonds just as occurs in many human cultures, and members of each sex also indulge in matings outside their pair bond. Much of what I have written about Common Guillemots has an anthropomorphic element to it, and indeed when watching them it is sometimes difficult not to credit them with human attributes. The higher incidence of extra-pair paternity among humans in the tenement block compared with the hospital sample in southern England (where there are few tenements) is also similar to birds. Among birds breeding in colonies the frequency of extra-pair copulation is considerably higher than among solitary breeding species, presumably because proximity increases the opportunity for such behaviour (Birkhead and Møller 1992).

* * *

What is it that makes someone want to study something as apparently obscure as the mating behaviour of birds? We might rephrase and broaden this to ask what is it that motivates scientists? There is no simple answer and no single factor that makes someone want to become a field biologist. The majority of my colleagues who are behavioural ecologists and field-workers were interested in natural history as children. For some reason many of my male associates, but few of my female colleagues who now study birds, were birdwatchers as children. Why should this difference between the sexes exist? I have discussed this with interested parties and three ideas emerged. First, parents may be happier about letting boys go off birdwatching on their own. Second, my female colleagues have suggested that boys may be more likely to be given and trusted with the tools of the trade, binoculars. I think this is an unlikely explanation, since many budding birdwatchers, myself included, endeavoured to watch birds long before being thought responsible enough to have binoculars. A third suggestion, also by my female friends, is that boys are more obsessive and competitive and may focus more narrowly than girls on some aspect of natural history. There may be additional explanations, but I think this last contains more than a grain of truth.

Plenty of people watch birds when young. What determines whether they remain birdwatchers, or manage to combine this interest with science and carve a career out of it? I think what makes the difference is that the budding scientist becomes dissatisfied with simply observing, and wants instead to find out why birds do certain things. Hence the title of a book on animal behaviour by Nobel prize winner Niko Tinbergen (1958) *Curious Naturalists*.

Wanting to know what makes things tick is certainly one of the things that motivates scientists, but there is slightly more to it than that. To appreciate why involves a slight digression. Trying to understand why an animal behaves this way or that requires the biologist to follow a fairly well defined set of rules. On seeing a piece of behaviour we do not understand—for example, a male Common Guillemot rushing over and forcibly copulating with his neighbour's partner—the initial step is to come up with some ideas that might sensibly explain this. The next step is to think up some observations or experiments that would disprove this hypothesis. If, after subjecting your hypothesis to most rigorous testing by observation, or better still by an experiment, it still looks good then that is probably the best you can ever do. You can be as sure as you can that your hypothesis is correct.

Having several hypotheses to explain a particular behaviour is far better than just having one. Let us take a specific example already familiar from this chapter. You are sitting on a cliff-top watching Common Guillemots and you see a male forcibly copulate with another male's partner. After asking yourself 'why did that male behave in that way?' you come up with three hypotheses (you might well be able to think of others). The first one is simply that the male mistook the female for his own partner. The second is that the male was sick with some kind of hormone imbalance and was mating with any female he saw. The third hypothesis was that the male was not sick and knew very well that this was not his partner, but he wanted to try to fertilize a female in addition to his partner.

Once we have our hypotheses we need to sit down and think about exactly what we would predict if each of these was true. Having done that we would need to make the necessary observations or experiments to test each hypothesis.

If the first hypothesis, which we shall call the mistaken-identity hypothesis, is true we would predict that males would only perform copulations with other females if their own partner was not at the colony. A few hours of observation at the right time of the season would show that, contrary to the predictions made by this hypothesis, males whose partners were present at the colony did indeed perform extra-pair copulations. We can thus discount this hypothesis.

The second hypothesis, the sick-oversexed male hypothesis, predicts that relatively few males should be trying to copulate outside their pair bond. After all, if the male is sick it is unlikely that many individuals in the colony would be suffering from the same complaint. A second prediction is that an oversexed male should not be too fussy about which females he tries to copulate with. He might, for example, try to copulate with any female regardless of what stage she is at in her reproductive cycle. Detailed observation would show, however, that the majority of males in the colony at one time or another attempt to copulate with other males' partners. Moreover, they are very specific about when they do it (see below), so these two bits of information together indicate that hypothesis number two is pretty unlikely.

Our third hypothesis, the extra-pair fertilization hypothesis, predicts that males should attempt to perform their extra-pair copulations with females' that are most fertile. In other words they should not bother trying to copulate with females that have already laid their egg. This would be a waste of time since her egg has already been fertilized. In fact, as I have mentioned, males time their extra-pair copulations very precisely so that they occur a few days before the female lays her egg—exactly when she is most fertile.

This leads us back to the question about what it is that motivates biologists. Curiosity is a vital spark, but lots of non-scientists are curious about animal behaviour. What sets the scientist apart is the combination of curiosity and thinking up novel hypotheses to explain a particular behaviour. The emphasis here is on the word 'novel'. Having a new idea, a fresh way of looking at problems, is what gives scientists their biggest 'buzz'. Being the first to think of something and the first to do it, and the first to publish the results of such a study is what it is all about. This is because in science, unlike art, there is usually only one correct answer, and the person who gets there first is assured of some degree of fame. As Peter Medawar (1982) said '. . . what X misses today Y will surely hit upon tomorrow (or maybe the day after tomorrow). Much of a scientist's pride and sense of accomplishment turns therefore on being the first to do something—upon being the man [sic] who did actually speed up or redirect the flow of thought . . .'.

For example, prior to about 1970 few people explicitly thought about animal behaviour from an evolutionary standpoint. Although many field biologists studying animal behaviour used the term 'evolution' in their interpretation of behaviours, they did not use it in the explicit 'selfish-gene' manner which is now almost taken for granted. Geoff Parker of Liverpool University was among the first to view insect reproductive behaviour in this light. To put it another way, his hypotheses to explain the behaviours he saw dungflies performing, were evolutionary hypotheses. This amounts to saying: a particular male insect behaves in this way because by doing so he would leave more descendants than if he behaved in another way. When Parker's study of sperm competition in insects was published in 1970 the idea of looking at behaviour from an evolutionary viewpoint was entirely novel. Since this approach has also turned out to be an extremely productive way of examining animal behaviour, Geoff Parker's name lives on in the behavioural ecology literature. You cannot look at a paper on sperm competition in animals without coming across a reference to that original paper. So, having a good idea is one of the major motivating forces for scientists.

Having ideas is not easy, however. By definition almost, good ideas are few and far between. Immanuel Kant has referred to the business of doing research as 'the restless endeavour'. Nothing describes the process better. You cannot simply sit down and decide that you are going to come up with a brilliant new hypothesis during office hours. The image of the absent-minded professor, bound up in his own thoughts might give the impression that it is possible deliberately to think novel thoughts. However, the reality is usually rather different. Professors often are absent minded and scientists often are wrapped up in their own thoughts, but good ideas often come at unexpected moments: in the bath, in bed, or wherever. There are a number of things that one can do to help the process along, and doing anything but science is one of them. Walking is a great way to allow your mind to sort itself out. Darwin knew this and built for himself his 'sandwalk' in the wood at the end of his garden. The 'restless endeavour' Kant refers to is the process of keeping (lesser) ideas moving round in your head, either consciously or unconsciously, in the hope that they might 'gel' into something useful. When they do, there is a moment of exhilaration. Depending upon how big is the problem you have resolved the exhilaration may last seconds or weeks.

I may have given the impression here that all scientists are hoping some 'big' idea might come to them, but this is not true. Any problem in biology requires a number of ideas which form a pyramid, with one major, unifying idea at the pinnacle, supported by numerous lesser ideas. To take an example close to my own interests: at the time, the idea of looking at sperm competition in birds was a novel one. But having had that idea I then needed lots of further ideas to test the various predictions that the sperm competition idea generated.

If the main trunk of motivation is a blend of curiosity and novel hypotheses, there are also numerous side branches. Undoubtedly part of my motivation for studying seabirds has been that I enjoy the places where they occur, and Arctic regions in particular have a special attraction. However, where field studies are conducted has to be balanced against the place's suitability for obtaining information. At an extreme level there are some seabird colonies in the Arctic whose topographical settings are surrealistically beautiful, but they are useless for studying birds. At the other extreme, one of my colleagues has spent every spring and summer for the last 10 years in the botanical gardens of his university town, simply because the birds (not seabirds) there are relatively tame and easy to observe.

One of the motivational elements common to all scientists is the inner feeling of satisfaction when things are going right during the process of data gathering. With my study of Common Guillemot extra-pair copulation I knew that my idea was good because at that stage few others had thought to see whether Geoff Parker's sperm competition ideas could explain some of the sexual behaviour of birds. I also knew that the fieldwork was feasible, but during the time I was watching the birds and collecting data I got tremendous satisfaction from knowing that all was going according to plan. This satisfaction also extends into the future because at this stage one can see that it will be possible to write an acceptable scientific paper based on the results.

After the fieldwork is over there follow a number of stages, some of which can also be gratifying. The carefully collected information must be sorted and synthesized in order to obtain a general picture. One cannot simply recount a succession of anecdotes about the birds' behaviour, rather one must try to examine the overall picture without losing sight of the individual differences. Providing the questions one asked at the outset were the right ones, the analysis of the results is relatively straightforward. For example, one question I asked myself was whether the relative number of sexually active males at the colony had any bearing on the likelihood that a female would be subject to an extra-pair copulation when she returned to the colony. Here was a clear question with a clear prediction. If my hypothesis was correct then the more biased the sex ratio was towards males the more likely it would be that any female would be subject to an extra-pair copulation. If the hypothesis was wrong, no such relationship would exist. Fortunately, in this case my hypothesis was correct, so writing about this was both simple and satisfying. With every emerging result that supports one's hypothesis, the satisfaction grows. Obviously, unless one is remarkably lucky or skilful, not every bit of data will fit exactly. The process of writing a scientific paper often triggers new ideas, and when this happens it is easy to wish you had collected some different or additional information. Occasionally though, you find that you already have the required information, collected for some other reason. If there's enough of it and it supports your ideas, that is fine. If the data are too few, then usually one can do no more than accept that however well thought-out the project was, it is impossible to think of everything.

Finishing the paper is also a good feeling. This is usually a sense of relief rather than elation. In my case at least, after having been through several successive drafts on the word processor, I am very pleased to be able to say to myself 'that's it—finished'. The elation comes later. A small dose first when packing up the manuscript to send it off to the editor of a scientific journal. A larger dose several months later if the manuscript is accepted for publication by the journal. I suppose the moment of acceptance is a particularly satisfying part for most scientists. Finally seeing the paper in print, usually 1 or 2 years later, produces merely a warm glow. My diminished enthusiasm for this final phase of the scientific process results from the time interval between completing the fieldwork and seeing the paper in print. The delay is inevitable because

of editing and printers and so on, and by the time the paper finally appears I am busy thinking about a new set of problems, or even engaged in fieldwork on a new topic or species. However, even if familiarity breeds a little contempt during the long wait, if what you have done is both well done and breaks new ground then the latter stages of the process are that much more satisfying.

There can also be a down side to all this, but you do not hear much about it. Scientists rarely publish studies that go wrong, and space in the Journal of Negative Results is often limited. Sometimes you can have what you consider to be a great idea and after checking it and double checking it you launch into a programme of research, only to find half way through that it was not such a great idea after all. Then you are left with the dilemma of whether to abandon the project or attempt to salvage something from it. In the latter case the result is often mediocre, with all that that entails.

The way science is run at present means one has to do everything possible to avoid things going wrong. Those who cannot do so, or who are unlucky are weeded out by the funding process. The rules for who gets and who does not get funding are increasingly stringent. Entire university departments are under scrutiny: are they publishing enough papers? This is fine, up to a point. Scientists in universities have got to earn their living, but the current system demotivates people and, worse, probably encourages safe, but boring science. If you know you have got to produce a certain number of papers each year to ensure your department gets a good rating, the last thing you are going to do is try anything too speculative. Yet, this is precisely how progress in science occurs.

CHAPTER 9

Changes

It is the naturalist's privilege to choose almost any kind of plant or animal for examination and be able to commence productive work within a relatively short time.

Wilson (1987)

After an interval of 9 years I returned to Labrador in 1992. This year was also significant in other respects, some positive, some negative. On the plus side, Prince Leopold Island in the High Arctic (see page 22) was finally given legal protection by being declared a National Monument—a mere 20 years after its unique biological nature had been identified. On the down side, 1992 marked the indefinite closure of Canada's 300-year-old cod fishery because there were simply no more fish left to catch. The demise of this once enormous fishery marks another notch in the ratchet of our self-destruction, and a notch that was barely audible outside Canada.

Just 2 days after leaving Sheffield I was back on the Gannet Clusters, making my way across the shell-beach on GC2 and up towards our cabin. Having been warned by people in Cartwright that our accommodation had not withstood the test of time too well, I was anxious to see exactly what we had to contend with. As I arrived, the door was blowing back and forth in the strong breeze, and the sight of the mess inside took my breath away. Fishermen had taken anything worth taking, the weather had ripped holes in the roof and a fox had used the cabin as a winter residence after the door had been left open. The result was a wet, stinking and apparently uninhabitable mess. My stomach churned with clashing emotions—I was excited about returning to

Inside the GC2 cabin as it was when we arrived in 1992.

the Gannet Clusters, but depressed and angry at the violation of the cabin. At the same time my feelings towards the fishermen were mixed. Initially it looked as though the fishermen had actually vandalized the cabin, but in fact all they had done was to open some large containers with an axe and then steal our chairs, fridge, cooker, stove and tools. There was no malicious damage. Most of the destruction was a direct result of ice, wind and rain, and the fox which had chewed everything chewable, and had left his unambiguous foxy leaving cards everywhere (literally, there were fox droppings on all the bunks and on every shelf, even those up in the roof space!).

After a couple of days habituation and reflection I viewed things more philosophically: because we had not been back to the Gannet Clusters for several years it probably was not unreasonable for fishermen to assume we were never coming back, and to feel justified in using whatever we had left behind. Life on the Labrador coast is hard in a declining fishery, and I could not actually begrudge them our equipment. It was unfortunate however, that they had left the door open after their visit.

There were four of us, David Nettleship, Gary Glenn, Gary Burness and myself, and we had come back to see whether changes in the fishery had affected the birds. As well as no cod, further south in Newfoundland there were no capelin either and that fishery too had been closed (see also page 75). When we landed on the Gannet Clusters a storm was brewing and, as our transport, the M.V. *Blue Thunder*, scuttled back to Cartwright's protected harbour, we had two choices: either to live in the tent which we had brought with us, or to clean up the cabin and use that. With heavy drops of rain and increasing wind, the cabin seemed unexpectedly inviting. The two Garys set about fixing the roof while David and I started on the inside. Remarkably, within just a few hours the cabin began to look like a home again and I was treated to a flood of memories as I cleared up everything from the floor: a familiar plate, a favourite book (now wet and sadly swollen), and our frisby! By the time it was dark the roof was covered in tarps, and the mess of rope, broken glass, soggy paper and mud had largely gone from the cabin floor. Outside, a fierce gale raged, and we breathed a sigh of relief that we had opted not to use the wildly flapping tent.

The storm continued at force 10 throughout the next day, so it was impossible to start our studies. Instead, we continued trying to get the cabin as clean and waterproof as possible. There was still a deep layer of 'mud' all over the floor, and there were numerous enthusiastic leaks in the roof and walls which needed to be fixed. By the evening the cabin was really habitable and since the rain had stopped I took the opportunity to climb up to the top of the island behind the cabin. Clambering over the rocks towards one of the Razorbill colonies I was amazed to flush a fox from beneath me. The fox ran off at high speed, and I got little more than a glimpse of it. Looking down to where it had been, I saw a blood-stained Puffin lying on the vegetation. It was still warm and when I picked it up the blood from its breast and neck continued to flow over my hands. Carrying the bird with me I went after the predator, anxious to check whether it was a red or an arctic fox. Before leaving Cartwright I had spoken to Bill Elson, who had told me that red foxes were particularly abundant at present, apparently because there was no market for skins, no-one trapped them any longer, with the result that the number of foxes had increased. He told also me that arctic foxes regularly came down from the north on the ice in the winter, but they usually returned before the break-up of the sea-ice in spring.

A few hundred metres away the fox stopped among the lichen-covered rocks to look at me. Through my binoculars I could see the small ears and soft brown pelage that made this an Arctic Fox. As we stared at each other the implications of a fox being on the island during the birds' breeding season started to dawn on me. After a brief pause the fox turned and disappeared among the rocks. By that time I had virtually reached the Razorbill colony, and as I made my way forward I could see that, like the Puffin I was holding, the colony was utterly dead. On my last visit here 9 years before this had been a bustling, thriving colony with dozens of birds sitting around on whitewashed boulders and with eggs or chicks in every crevice, but now there was nothing. The rocks were totally devoid of droppings and it looked as though few birds had been anywhere near the colony, let alone attempted to breed.

I retraced my steps and then climbed rapidly to the other Razorbill colonies, only to discover exactly the same ghost towns. Overall, it looked as though the entire population of 1200 pairs of Razorbills on GC2 had failed to breed. These represented about 20% of 5800 pairs on the Gannet Clusters as a whole—presumably because of the fox. I then checked the small Common and Brünnich's Guillemot colonies, only to find the same effect. Finally, as I made my way back to camp I picked up the chewed wings of another fox victim—a Leach's Petrel.

Back at the cabin we discussed the effect of this unwelcome predator on the island and what we should do about it. We decided that since 'my' guillemots and the Razorbills had suffered a total breeding failure it was too late to do anything for them this season. On the other hand David was optimistic that 'his' Puffins would be able to cope. After all, he said, Puffins and red foxes had coexisted on Baccalieu Island off Newfoundland for several years.

Our next discovery did little to encourage the idea of peaceful coexistence. Behind the cabin I noticed some freshly dug peat, and on peering into an enlarged Puffin burrow I found a dead Puffin. I pulled it out, only to see several more lying deeper inside the burrow. I unearthed them and by the time I had finished I had recovered no less than 20 Puffins, one Common Guillemot and a Razorbill. This was a fox cache—a hidden supply of food for future needs, and indisputable evidence that coexistence was not the name of the game.

As so often happened on the Gannet Clusters, the day following the storm was bright and clear, with a rapidly calming sea. I was anxious to get across to GC4, our main guillemot study area to see how both the small cabin and the birds there had fared over the past decade. By mid-morning the swell had subsided sufficiently for me to feel confident about being able to land on GC4. To my relief the cabin there, where I had previously spent so much of my time, was intact, dry and untouched. This was remarkable given that I knew people from Cartwright had been here and had looked inside. More than anything else, being here brought back a rush of memories: everything was much as we had left it, and it felt as though I had barely been away. The cabin smelt the same, and our books and papers were still on the desk. The only casualties were the tins of food which had badly rusted and did not look too appealing, but some jars of jam and marmalade which I opened were perfect. I checked the 'sell-by' or 'best-before' dates on these, only to find that most of them actually fell before the time we had originally purchased them in Cartwright!

Outside the cabin, along the edge of the guillemot cove our hides were all still in place and were in good shape. One had lost a door, but otherwise they had withstood the elements extremely well. I prised open the door to my hide, and peered anxiously through the window to see what had become of my guillemots. There were plenty of birds on the Common Guillemot study plot below my hide, but there were also some changes: a huge boulder had slipped across the plot and had altered the configuration of the breeding colony. Rather surprisingly I could see no Common Guillemot chicks, nor any indication that any chicks had hatched. As I counted the birds I also noticed quite a few sites occupied by birds without an egg or chick, and that the few eggs I did see all looked remarkably clean, suggesting that they might have been laid relatively recently. This was odd—it was late July, and even in our latest season (1982) most birds had got chicks by this date. However, I knew that it had been a very cold spring in eastern Canada, so I assumed that what I was seeing was the result of a very late breeding season.

In complete contrast the situation for Brünnich's Guillemots was remarkable—their numbers had increased enormously, and opposite my hide there was an entirely new colony! In fact, when I later counted all the Brünnich's Guillemots at the Gannet Clusters I found 2025 individuals—over twice the number we had recorded in 1983 and continuing a trend which we had previously suspected (see Fig. 13). Here at last was something encou-

Brünnich's Guillemot.

raging. Also, rather to my surprise I noticed that one or two of the Brünnich's Guillemot were brooding chicks. This too was unexpected because in our earlier study Brünnich's Guillemots always bred slightly later than the Common Guillemots, yet here they were with chicks while the Common Guillemots were still on eggs.

After spending several hours watching and counting guillemots on GC4, I checked the Razorbills, only to find that their breeding numbers were down, and in one or two cases, entire colonies had been abandoned. Some Razorbills were incubating eggs, but I saw no chicks, and I began to wonder if there might be a fox on GC4. This was easy enough to check, GC4 is a tiny island (page 182), and on walking around it I saw no fox, nor any footprints in the few muddy pools. It is often said that going back to a place where you have had a really good time can be disappointing. It is difficult to recreate the past: at worst it fails to fulfil one's expectations, and at best nothing has changed. Here though, so many changes, some good, some bad, promised an intriguing biological story to unravel.

The summer of 1992 was a cool one along the Labrador coast. When we arrived at the Gannet Clusters in late July icebergs were abundant and on the islands many of the spring flowers were only just beginning to bloom. The heath-like vegetation around our camp on GC2 was brightened by brilliant purple beach peas, white bunchberry and clusters of star-like stitchworts. As always, birds other than seabirds were few and far between, but on several occasions we were treated to the spectacle of an immature Peregrine Falcon swooping after Puffins and Kittiwakes over the islands. After a storm, a flock of 60 Phalaropes, both Red-necked and Grey, foraged furiously in a tide-rip off the end of GC2. Several species of waders (shorebirds) passed overhead, including Greater Yellowlegs and White-rumped Sandpipers, and it was uncanny to think that they had completed their breeding season further north while many of the seabirds on the Gannet Clusters were still incubating eggs. Passerines were especially scarce, but one morning I discovered a Red-breasted Nuthatch which, in the absence of any trees, was foraging over the rocks close to the cabin. Our most spectacular visitor was an immature Bald Eagle, which roosted for several nights on the cliffs on the north side of GC2. Remarkably, the only birds to acknowledge the eagle's presence were the Great Black-backed Gulls, who mobbed it. Even when the eagle flew over their breeding colonies the auks and Kittiwakes simply ignored it. I made another surprising discovery: the day after I had left an apple core in the hide on GC4 I noticed it had been moved. I wondered whether there were still mice on the island. To check this I smeared a mixture of peanut butter and popping corn onto a flat board and left it in the hide overnight. The next day I was delighted to find all the food gone, but the board liberally dotted with mouse droppings. Contrary to my earlier conclusions (see Chapter 7) the Deer Mouse population had persisted despite a weasel and other predators!

At the main camp at GC2 a movement on the hill behind the cabin caught my eye, and to my amazement I saw not one, but four foxes romping around on a small crowberry plateau! There were two adults and two cubs. On seeing them our perspective on the fox situation lurched to a new level. Initially it had seemed like an unlucky (for the birds), chance event that had left one fox behind after the winter ice had broken up, but what we were dealing with now was a breeding population—albeit a small one. Their discovery also raised the possibility that there were foxes on some of the other islands and led to a discussion similar to the one we had had over the weasel on GC4 in 1983. Unlike that previous situation, letting nature take its course now seemed a much less sensible option. The Gannet Islands constitute the single most important seabird colony in Labrador (Nettleship in press), and although the fox invasion was a natural event, it was also probably an extremely rare one. George Cartwright (1792) hardly ever encountered arctic foxes after the ice break-up during his time on the Labrador coast, and neither we, nor any of the other ornithologists that had visited the Gannet Islands in more recent times, had ever recorded foxes on these offshore seabird islands. From what we had seen already, the damage that one small family of foxes could do to a population of seabirds was truly remarkable.

Over the next few days we systematically visited the other islands to check for signs of foxes. Our small aluminium boat was not designed for adventure, so we did not travel as far as GC6 or to Outer Gannet (see Fig. 12)—instead we confined our attention to the five Gannet Clusters. On GC5 we found two more foxes, but no more pups. One at least was a male since I got close enough to him to be able to see his reproductive equipment. His confiding nature was enchanting, but that only made our dilemma worse. I discovered this fox curled up asleep in the sunshine, lying on a rock encrusted with huge black leafy lichens. As I crept towards him, hardly daring to breathe, he sleepily raised an ear and opened an eye, but did not move away until I was a few metres away from him. Part of me was thrilled by being able to watch such a beautiful animal at close quarters, while another part of me wanted to see him and his fellows removed from this fragile seabird colony. The effect of the two foxes on GC5 was identical to that on GC2—all Razorbill colonies had been abandoned, there were no Common Guillemots breeding, no Canada Geese. Even the White-crowned Sparrows, and the beautiful Horned Larks which had been such a feature of our earlier trips were absent, and the islands were sadly silent without them. The crowberry and willow heath on GC5 was littered with the carcasses of adult Puffins and their eggs, and we even found the wings of some Leach's Petrels which we had not known previously to breed on GC5. The only seabirds I found breeding on GC5 were two Razorbills incubating their egg on tiny ledges on a steep cliff at one end of the island. Presumably this was the only place where the foxes could not reach them.

Seeing these two Razorbills made me think back to what I had seen on GC4, where the cliff-nesting Brünnich's Guillemots had chicks, but the Common Guillemots, which nest in areas much more accessible to foxes, were still incubating their eggs. I wondered whether foxes had actually been on that island earlier in the season despite their now obvious absence.

Before I had a chance to check this, David and I made a visit to GC3, the

Iceberg lying between GC3 and the mainland.

island immediately to the south of GC2. This is a very low-lying island with huge numbers of Common Guillemots (about 15,000 pairs), and one where the birds are very vulnerable to human (or any other) disturbance. For that reason GC3 was always one of the Gannet Clusters we visited least. However, I had been there on my very first visit in 1978 and, by crawling along on my stomach, had managed to photograph some of the Common Guillemot breeding areas without causing any disturbance. Our current visit to GC3 was to find out how the Razorbills had fared compared with those on the other islands. Just as on GC4 we found that the Razorbills did not seem to be doing very well: only about 20% of breeding sites were currently in use and contained an egg. It was quite clear that there were no foxes present on this island, although we were not sure whether they had been there earlier in the season.

After we had finished checking the Razorbill sites I crept up to look at a Common Guillemot area I had photographed in 1978. As I peered cautiously over the edge of the rocks into the colony I was confronted with one of the most devastating sights I have ever seen in a seabird colony. Where there had once been several thousand birds there was now nothing but a sea of abandoned eggs. I stared down in disbelief, unable to comprehend what could possibly have caused such mayhem. There were eggs everywhere, some whole, others smashed from having rolled off their ledges, and others had evidently had their contents eaten by Great Black-backed Gulls. In some areas eggs had rolled into piles half a metre deep. What was so disturbing was that whatever had caused this damage had forced the majority of birds to abandon this breeding area altogether. I called David over and we climbed down into the colony to look at the damage more closely. We counted no fewer than 2000 abandoned eggs, and the rather pathetic spectacle of a handful of isolated adult Common Guillemots which, rather surprisingly, were brooding newly hatched chicks.

Still feeling somewhat shocked I returned to GC4 to see if I could find any evidence that foxes had been there, which would then explain why so few Razorbills were breeding and why the Common Guillemots' breeding season was so delayed. As I landed and met up with one of the Garys, he told me he had found a guillemot egg hidden in the long grass by the cabin. If it had been cached by a fox, this would provide just the evidence needed to prove that foxes had indeed been on GC4 earlier in the season.

Food hoarding is widespread among birds and mammals and whenever animals such as foxes, magpies and so on hide food, they do so very carefully to prevent others from finding their cache. The guillemot egg that Gary had found had been well hidden, but it had been exposed when the heavy rain had washed away part of the bank. I was fairly sure this was a fox cache and I began to search in other similar places along the bank, concentrating particularly on those areas at the base of grassy clumps. To my surprise I found several more eggs, of both Common Guillemots and Puffins. Later I scoured the grassy surface of the island and found several old fox droppings, thus confirming that one or more foxes had been on GC4.

The information we had gathered so far from the various islands was difficult to interpret and our situation reminded me of the sort of problem archaeologists face when trying to construct a vase from shards of pottery. For those islands where we knew foxes to be present the evidence was straightforward, but on the other two islands, GC3 and GC4, it was more equivocal. It looked as though foxes had been present on GC4 at least until the time when Common Guillemots and Puffins started to lay eggs, but they had probably been on GC3 more recently judging from the Guillemot mayhem we had seen.

The final island we checked was GC1. We could see most of this island from our camp on GC2 and we were certain there were no foxes present. However, we still needed to check whether they had been there. So, in an unpleasant surging swell we landed precariously on the northern end of the island and made our way towards two Razorbill study areas we had established in 1983. Before we even started looking in detail at the colony it was apparent that, compared with the other areas, this was a very healthy colony. There were plenty of birds and virtually every nest-site contained either an egg or a chick. Overall, we found there were slightly more occupied breeding sites than there had been in 1983, suggesting there had been no fox disturbance here. When we moved on to check the Common and Brünnich's Guillemots the same was true for them too: there were plenty of eggs and chicks, and everything was much as it should have been. We were relieved to find that on one of the five islands things seemed to be relatively normal. We searched in vain for cached food or fox droppings, further confirming that for some reason the foxes had not got here during the present breeding season.

One more question needed answering to double check our findings, and that was for Puffins. On the two islands where the foxes were living, GC2 and GC5, relatively few Puffin burrows contained an egg or chick. This was hardly surprising since the foxes spent much of their time hunting through the Puffin colony, causing tremendous panic flights. The foxes were also sufficiently small to get their head and much of their body down the Puffin burrows, enabling them to capture adults and take their eggs. We saw foxes take and cache a large number of Puffin eggs, and we also found the remains of dozens of eggs which the foxes had eaten. At the time of our visit only 18% of 108 Puffin burrows on GC2 contained eggs, and the corresponding figure for GC5 was similar: 25% of 117 burrows. It seems likely that by the end of the birds' breeding season no Puffin chicks would be reared successfully on these islands. If we were right about GC1 not being visited by foxes then Puffin breeding success there should be much higher.

In due course, and in pouring rain, David checked 176 Puffin burrows on GC1, confirming our suspicions by finding that 89% of burrows contained an adult Puffin brooding either an egg or a chick!

With this final piece of information we were able to reconstruct a likely sequence of events. The foxes had undoubtedly reached the Gannet Islands via the winter pack-ice but for them to remain into the birds' breeding season must be extremely unusual. The low-lying nature of these islands means that most of the breeding seabirds are easily accessible and regular visitations by foxes would have certainly eliminated all birds that were not breeding on cliff ledges. In their now classic *Sea-birds*, Fisher and Lockley (1954) wrote: 'It is a commonplace that in Spitsbergen and Greenland, and no doubt in other parts

Arctic fox resting outside its den on GC2.

Arctic fox carrying a Puffin egg which it is about to hide.

of the Arctic, the presence of arctic foxes forces the seabirds to occupy the more inaccessible ledges of the cliffs'. In the Aleutian islands, Alaska, when arctic foxes were introduced (for fur farming) between 1900 and 1936 their effect on ground- and burrow-nesting seabirds was, inevitably, catastrophic (Sekora *et al.* 1979). Our experience in Labrador in 1992 was probably unique in that we were able to witness the process in progress. The situation at the Gannet Clusters was also unique in other respects: first it was an entirely natural, if rare event, and second, the presence and absence of foxes on the five adjacent islands provided a beautiful natural experiment, which allowed us to demonstrate the effects of the foxes unequivocally. Our attempts to summarize and reconstruct the events in 1992 are shown in Fig. 25.

Our main conclusions were that on the two islands (GC2 and GC5) with resident foxes, none of the open-nesting species, like the Razorbill or Common Guillemot bred successfully. We knew that at least some individuals of these species had attempted to breed because we found a few cached eggs. However, I suspect that when the first birds to lay became victims of the foxes or had their eggs taken the others simply remained away from the colony and probably failed to produce eggs. The birds from these abandoned colonies simply sat disconsolately on the sea beside their breeding area, only occasionally daring to venture any closer. This was particularly noticeable among the Razorbills, and at dusk I could walk up to the top of GC2 and see a raft of birds sitting immediately opposite each of the six main colonies. The foxes may have been on GC2 for more than a year, and I wondered how long a Razorbill or Common Guillemot might wait before deciding to try and find a new site somewhere else.

At the other extreme there was no evidence that foxes had even visited GC1, and everything there appeared to be normal. This island was our 'control' against which we could compare all the other islands. Although there was a full complement of birds, all of whom were breeding, our observations on GC1 showed that the 1992 breeding season was about 1 week later than we had previously recorded (Birkhead and Nettleship 1987a)—presumably an effect of the cold spring and summer.

On the remaining two islands where there had been foxes for at least a short time prior to our visit, their effect was intermediate, but left us with a number of uncertainties. The presence of fox scats and cached eggs on GC4 made it clear that foxes had been there and the relatively low numbers of breeding Razorbills was probably also a consequence of foxes. As on GC2 and GC5, my guess is that after seeing foxes, many Razorbills on GC4 just gave up the idea of breeding that year. The situation among the guillemots was more interesting. I checked the Common Guillemots on my study plot every day until we left in mid-August and none of their eggs had hatched. Yet, on the undisturbed GC1, almost all Common Guillemot chicks had hatched by this time. I think what happened on GC4 was that while the fox was present early in the season the Common Guillemots were put off egg-laying, and not until the fox had disappeared did they begin to lay, albeit extraordinarily late. Knowing that all birds on my study plot had laid when I started observations, and that none had hatched by the time we left the Gannet Clusters 2 weeks later, I

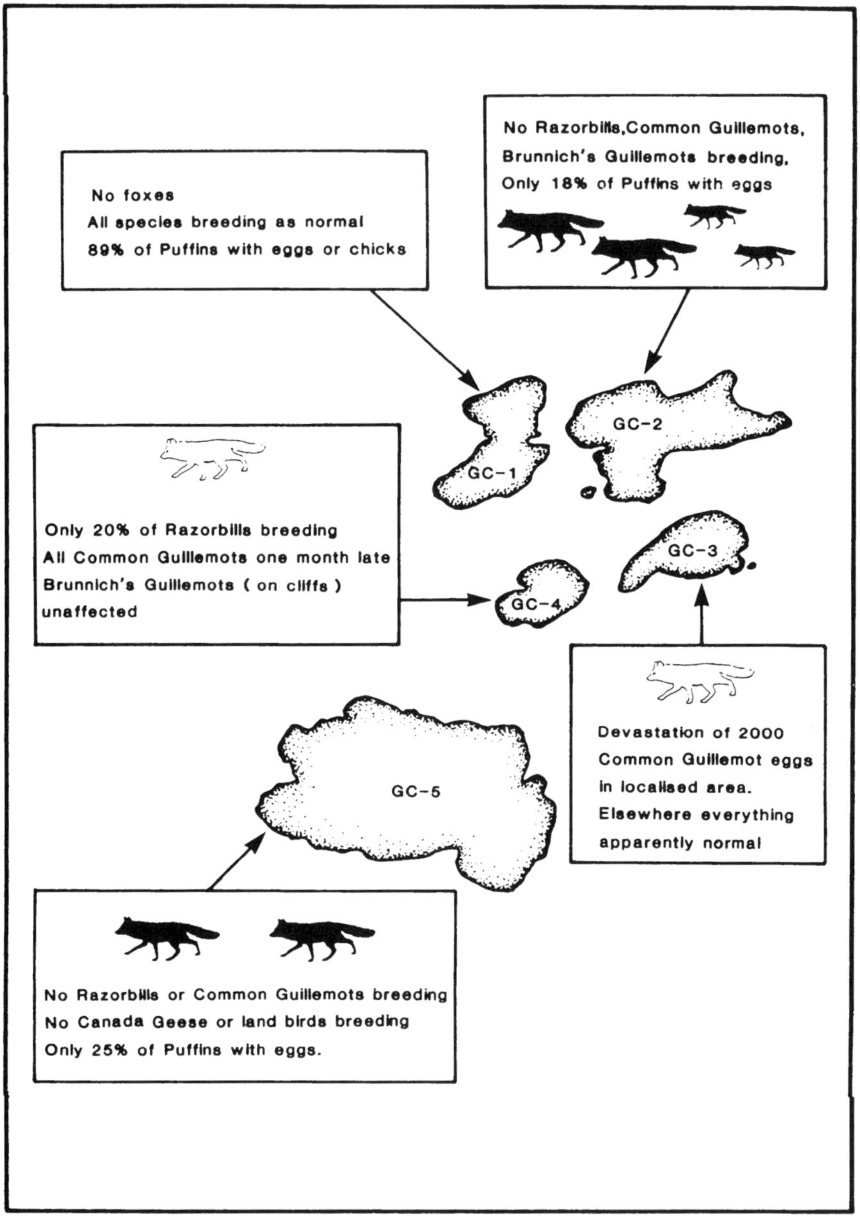

Fig. 25 *Summary of the effect of Arctic Foxes on seabirds on the Gannet Clusters in 1992. Islands with resident foxes are indicated by dark fox symbols (small symbols indicate cubs). Islands known (GC4) or suspected (GC3) of having visits by foxes are indicated by a light fox symbol. Island numbers are indicated.*

reckoned that the peak of hatching would occur somewhere between the 19 and 30 August—almost a month later than we had recorded in any previous year. What effect this late hatching would have on the young birds I do not know, but studies elsewhere have shown that the later in the season a Common Guillemot chick is reared, the less likely it is to survive.

Interestingly, the Brünnich's Guillemots on GC4 seemed to have been relatively unaffected by the foxes. The hatching of their young occurred at much the same time as their counterparts on the undisturbed GC1—presumably because their cliff-ledge breeding sites kept them safe from the foxes. I also think that the Brünnich's Guillemot's phlegmatic nature might have helped, and prevented them from panicking if a fox came near. Earlier I described how in our previous studies we found that Brünnich's Guillemots had relatively low breeding success compared with Common Guillemots and that this was a direct result of the type of breeding habitat they used (see page 193). Yet, here were Brünnich's Guillemots increasing in numbers, and to all intents and purposes immune from the foxes. The advantage of using this otherwise poor habitat may occasionally be considerable!

One of the unanswered questions regarding GC4 was: what happened to the fox or foxes? Judging from the timing of the Common Guillemot's breeding season there had not been a fox on GC4 for at least 2 weeks prior to our arrival in late July. In addition, I opened up the Common Guillemot eggs that the foxes had cached and found all of them to be unincubated, suggesting (from the evidence on GC1), that foxes had been present on GC4 at the very start of egg-laying in late June. In other words the foxes must have left GC4 sometime between late June and early or mid July. Whether they swam or whether there was ice which allowed them to walk off, or whether the GC4 fox died, we simply do not know.

The sequence of events which caused the Common Guillemot mayhem on GC3 is a mystery. We found no incontrovertible evidence of foxes on this island—neither scats nor cached food. We did find several dead adult Razorbills, however, which is unusual and suggests predation by something. After we had discovered the devastation in that one small cove, I went back and looked elsewhere on the island at Common Guillemot and Razorbill breeding areas, only to find that all was normal, and much as it was on GC1. In other words, the damage was extremely localized. I tried to estimate just when it had occurred by opening up some of the abandoned eggs. In each case the embryos were about 2 or 3 cm in length—about half way through their development. Assuming that the Common Guillemots on GC3 laid at the same time as the undisturbed birds on GC1 (i.e. in late June), the devastation probably occurred around the middle of July. I suspect also that whatever had taken place did so on just one day, but was sufficiently frightening that few birds dared to return after it was over. The cove where the incident took place was virtually enclosed and had a fox got in there the birds would have panicked in their efforts to find an escape route. Indeed, the way the eggs had poured into gullies and crevices suggested that there had indeed been a terrible stampede of birds, frantic to find a way out. A possible scenario is that a fox swam across from GC2, took some eggs, killed some adult Common Guillemots and

The south-western tip of GC3 – once a thriving Common Guillemot colony – destroyed, probably by an arctic fox, in 1992.

Razorbills, and then returned to GC2. The cove on GC3 was almost the closest point to GC2, so this is just possible. It is also possible that an exceptional piece of ice wedged itself between the two, islands allowing the fox access to GC3. Unfortunately we shall probably never know what took place. We considered but rejected the idea that the destruction we saw was an act of human vandalism, reminiscent of the 'egging' that took place on these and other islands in the not too distant past. However, had this been the case I am sure we would have found some signs of human presence. Moreover, there are many easier places on the Gannet Clusters to collect eggs or kill adult Common Guillemots than this particular cove.

At the end of the day we were left with a moral dilemma—what should one do about a natural predator, which by a mere fluke, turns up at an internationally important seabird colony? The effect of the foxes was considerable, and

Abandoned Common Guillemot eggs on GC3 in 1992.

overall we estimated that the foxes had caused 80% of all Razorbills and at least 25% of all Puffins to fail in their breeding in 1992. The effect on the other species was less marked and despite the scenes of Common Guillemot carnage, the 2000 abandoned eggs we found represented only 5% of those laid on the Gannet Clusters that year. However, the foxes' activities may not have been confined to a single breeding season. It seems likely that they had been on GC2 for at least the two previous seasons, and the fact that they were breeding there suggests that they may have even been the parents of the individuals on GC5. The implication was that eventually the other islands too could be permanently occupied. What was also remarkable was that a predator not much larger than a domestic cat could cause such destruction. In the event we did not interfere, but were content, for this season at least, simply to record what we observed. Subsequently, the Canadian Wildlife Service decided that the foxes should be removed. However, this venture was only partially successful, and as I write there are still three foxes present on the Gannet Islands.

* * *

We went to the Gannet Clusters and found a natural ecological disaster, although our original motivation for returning was in anticipation of a man-made problem. The closure of both the cod fishery and the capelin fishery in eastern Canada in 1992 meant that conditions in the marine environment had reached an all-time low and we were anxious to see whether there had been any corresponding changes in the diet and feeding habits of the Gannet

Clusters' seabirds. Because breeding was so late in 1992 we were able to get information only for the two guillemot species. To our surprise and relief we found that the diet of these two species was virtually identical to that in our earlier studies (see page 207). Common Guillemots continued to feed their chicks largely on capelin, and Brünnich's Guillemots fed theirs mainly on daubed shannies and other bottom-living fish species. In this part of Labrador, at least, it appeared that despite a general scarcity of capelin, Common Guillemots were continuing to find them to feed their offspring. What we did not know was how much harder the birds were having to work to obtain this food, and we would only be able to determine this by additional detailed studies. However, our results confirm something that other seabird biologists have suspected for some time, namely that studies of marine birds often provide better information on what fish are available than do studies by fisheries biologists.

Herein lies the problem. The assessment of fish stocks is undoubtedly difficult, and predicting the effect of removing one species, such as the cod, from the system, is even more difficult. What is clear is that if you want the cod to recover, the last thing you do is open another fishery designed to destroy its main prey, the capelin. Nonetheless, this is exactly what happened (see page 75).

Canada's termination of its cod fishery marks the end of one of the world's greatest wildlife resources. But because cod spend their entire lives out of sight in the sea, the closure of its fishery caused little more than a ripple of concern, at least outside its eastern fishing communities. In reality the demise of the cod in eastern Canadian waters is akin to the ruthless exploitation of the American bison and the blue whale or the extirpation of the Great Auk.

APPENDIX I

List of common and systematic names

BIRDS

Adelie Penguin	*Pygoscelis adeliae*
American Pipit	*Anthus rubescens*
Ancient Murrelet	*Synthliboramphus antiquus*
Arctic Skua	*Stercorarius parsiticus*
Atlantic Puffin	*Fratercula arctica*
Bald Eagle	*Haliaeetus leucocephalus*
Baird's Sandpiper	*Calidris bairdii*
Black Guillemot	*Cepphus grylle*
Brünnich's Guillemot (Thick-billed Murre)	*Uria lomvia*
Californian Condor	*Gymnogyps californianus*
Cassin's Auklet	*Ptychoramphus aleuticus*
Chinstrap Penguin	*Pygoscelis antarctica*
Chough (red-billed)	*Pyrrhocorax pyrrhocorax*
Common Guillemot (Common Murre)	*Uria aalge*
Craveri's Murrelet	*Brachyramphus craveri*
Crested Auklet	*Aethia cristatella*
Eider (common) Duck	*Somateria mollissima*
Emperor Penguin	*Aptenodytes forsteri*
Erect-crested Penguin	*Eudyptes sclateri*
Fieldfare	*Turdus pilaris*
Fiordland Penguin	*Eudyptes pachyrhynchus*
Florida Scrub Jay	*Aphelocoma coerulescens*
Fulmar (Northern)	*Fulmarus glacialis*
Galapagos Penguin	*Spheniscus mendiculus*
Gannet (Northern)	*Sula bassana*
Gentoo Penguin	*Pygoscelis papua*
Glaucous Gull	*Larus hyperboreus*
Golden Plover	*Pluvialis apricaria*
Goshawk	*Accipiter gentilis*
Great Black-backed Gull	*Larus marinus*
Greater Yellowlegs	*Tringa melanoleuca*
Grey Phalarope	*Phalaropus fulicaria*
Gyrfalcon	*Falco rusticolus*
Harlequin Duck	*Histrionicus histrionicus*
Herring Dull	*Larus argentatus*
Horned Puffin	*Fratercula corniculata*
Humboldt Penguin	*Spheniscus humboldti*
Iceland Gull	*Larus glaucoides*
Jackass Penguin	*Spheniscus demersus*
Japanese Murrelet	*Synthliboramphus wumizusume*
King Eider	*Somateria spectabilis*

King Penguin	*Aptenodytes patagonicus*
Kittiwake (black-legged)	*Rissa tridactyla*
Kittlitz's Murrelet	*Brachyramphus brevirostris*
Knot	*Calidris canutus*
Lapland Bunting (Longspur)	*Calcarius lapponicus*
Leach's Petrel	*Oceanodroma leucorhoa*
Least Auklet	*Aethia pusilla*
Little Auk (Dovekie)	*Alle alle*
Little Penguin	*Eudyptula minor*
Long-tailed Skua	*Stercorarius longicaudus*
Macaroni Penguin	*Eudypytes chrysolophus*
Magellanic Penguin	*Spheniscus magellanicus*
Marbled Murrelet	*Brachyramphus marmoratus*
Parakeet Auklet	*Cyclorrhynchus psittacula*
Peregrine Falcon	*Falco peregrinus*
Pigeon Guillemot	*Cepphus columba*
Raven	*Corvus corax*
Razorbill	*Alca torda*
Red-breasted Nuthatch	*Sitta canadensis*
Red-necked Phalarope	*Phalaropus lobatus*
Red-throated diver	*Gavia stellata*
Rhinoceros Auklet	*Cerorhinca monocerata*
Rockhopper Penguin	*Eudyptes chrysocome*
Royal Penguin	*Eudyptes schlegeli*
Snares Island Penguin	*Eudyptes robustus*
Snow Bunting	*Plectrophenax nivalis*
Spectacled Guillemot	*Cepphus carbo*
Thayer's Gull	*Larus thayeri*
Tufted Puffin	*Lunda cirrhata*
Turkey	*Meleagris gallopavo*
Wandering Albatross	*Diomedea exulans*
Whiskered Auklet	*Aethia pygmaea*
White-crowned Sparrow	*Zonotrichia leucophrys*
Xantu's Murrelet	*Brachyramphus hypoleucus*
Yellow-eyed Penguin	*Megadyptes antipodes*
Zebra Finch	*Taeniopygia guttata*

MAMMALS

African Hunting Dog	*Lycaon pictus*
American Bison	*Bison bison*
Arctic Fox	*Alopex lagopus*
Arctic Hare	*Lepus arcticus*
Bank Vole	*Clethrionomys glareolus*
Beluga Whale	*Delphinapterus leucas*
Bluewhale	*Balaenoptera musculus*
Deer Mouse	*Peromyscus maniculatus*
Elephant (African)	*Loxodonta africana*
Grey Seal	*Halichoerus grypus*
Harp Seal	*Pagophilus groenlandicus*
Human	*Homo sapiens*

Humpback Whale	*Megaptera novaeangliae*
Minke Whale	*Balaenoptera acutorostrata*
Muskox	*Ovibos moschatus*
Narwhal	*Monodon monoceros*
Orang-utan	*Pongo pygmaeus*
Polar Bear	*Ursus maritimus*
Porpoise (Harbour)	*Phocoena phocoena*
Red Deer	*Cervus elaphus*
Red Fox	*Vulpes vulpes*
Short-tail Weasel (= Stoat)	*Mustela erminea*
Skomer Vole	*Clethrionomys glareolus skomerensis*
Spotted Hyena	*Crocuta crocuta*
Steller's Sea Cow	*Hyrdodamalis gigas*
Stoat (= Short-tail Weasel)	*Mustela erminea*
Walrus	*Odobenus rosmarus*
Weasel	*Mustela nivalis*

FISH

Bass (Striped)	*Morone saxatilis*
Capelin	*Mallotus villosus*
Cod (Arctic)	*Boreogadus saida*
Cod (Atlantic)	*Gadus morhua*
Daubed Shanny	*Lumpenus maculatus*
Fish Doctor	*Gymnelis viridis*
Lumpsucker	*Cyclopterus lumpus*
Menhaden (Atlantic)	*Brevoortia tyrannus*
Salmon (Atlantic)	*Salmo salar*
Sandeel (Sandlance)	*Ammodytes spp.*
Saury (Atlantic)	*Scomberesox saurus*
Shorthorn Sculpin	*Myxocephalus scorpius*
Snake Blenny	*Lumpenus lumpretaeformis*
Sprat	*Sprattus sprattus*

INVERTEBRATES

Easter tiger swallowtail	*Papilio glaucus*
Eastern black swallowtail	*Papilio polyxenes*
Tick	*Ixodes uriae*
Yellow dungfly	*Scatophaga stercorarius*

APPENDIX II

Notes on the local seabird names used in Labrador

TURR or MURRE (pronounced to rhyme with 'fur'). The Common Guillemot (Britain) or Common Murre (North America) *Uria aalge*. In Newfoundland and Labrador the names 'turr' and 'murre' are used indiscriminately and can refer to either Common or Brünnich's Guillemots. Most authorities (e.g. Lockwood 1984) suggest that, like the names Kittiwake and Cuckoo (and many other species), 'turr' and 'murre' are onomatopoeic and mimic the birds' calls. However, Potter and Sargent (1973) state that 'murre' has a Norse origin (stemming from the Latin *murmur*, meaning grumble, or growl), and that the original meaning of murre was 'growler'.

Not only is there current confusion regarding these vernacular names, there was also confusion, of a different sort, in the past. It seems clear that early observers were not very careful about distinguishing Guillemots from Razorbills—presumably because they are superficially similar and often occur at the same colonies. The result however, was that their early names were often interchangeable. Lockwood (1984) states that the term 'murre' was first recorded by Ray (1678) in Cornwall, meaning Razorbill, and was only used for the Common Guillemot in 1845 (Lockwood 1984). However, Cartwright (1792) refers to 'murrs' meaning Common Guillemots (Townsend 1911) (but only in Britain (at Portland Bill) and does not refer to them in Labrador).

TINKER: Razorbill *Alca torda*. This is derived from 'tinkershere', meaning a tinker's hue (i.e. black) and is a reference to the colour of the Razorbill's back (McAtee 1959). Lockwood (1984) states that this was originally a Dorset name for the Guillemot and it is recorded first in1799, but in fact Cartwright (1792) refers to 'tinkers', which Townsend (1911) states are Razorbills.

TICKLE-ASS or TICKLE-ACE: Black-legged Kittiwake *Rissa tridactyla*. The first part of the name refers to this species' habit of frequenting tickles (a Newfoundland and Labrador term for a narrow strait between the mainland and an island or between two islands). The meaning of the second part is obscure (McAtee 1959).

PIGEON or SEA PIGEON: Black guillemot *Cepphus grylle*. There are two reasons why this local term is used: the (apparently) pigeon-like shape of their bill, and their habit of laying two eggs, which they share with all pigeons and doves. Cartwright (1792) refers to Black Guillemots as sea-pigeons or pigeons.

BULL BIRD: Little Auk (Britain) or Dovekie (North America) *Alle alle*. Bull bird or bull apparently referred to this species' relatively short and bull-like neck. Cartwright (1792) refers to Little Auks as bulls. The name North American Dovekie is a diminutive form of 'dove', which the species was also thought to resemble (McAtee 1959). Lockwood (1984) states that Dovekie was also used in Scotland as early as 1821 for the Black Guillemot.

PARROT or SEA PARROT: Atlantic Puffin *Fratercula arctica*. A name inspired by the shape of the bill. The name sea parrot was first recorded in 1694 (Lockwood 1984), and this is how Cartwright (1792) refers to the Atlantic Puffin (Townsend 1911). The name 'puffin' originally referred to the Manx Shearwater *Puffinus puffinus* (see Lockwood 1984), but in 1764 Pennant used the name in the sense which we currently employ it.

NODDY: Northern Fulmar *Fulmarus glacialis*. Derived from the nodding movements of the bird's head in flight (McAtee 1959).

References

ANKER-NILSSON, T. & BARRETT, R. T. (1991). Status of seabirds in northern Norway. *British Birds*, 84: 329–341.

ANON. (1980). Alluring Labrador: a Journey Through Labrador. *Them Days Magazine*, Goose Bay, Labrador.

AUSTIN, O. L. JR. (1932). The Birds of Newfoundland Labrador. Memoirs of the Nuttall Ornithological Club. No. VII. Cambridge, Mass.

BARLEY, N. (1988). *Not a Hazardous Sport*. Viking Press, London.

BÉDARD, J. (1969). Adaptive radiation in Alcidae, *Ibis* 111: 189–198.

BELOPOL'SKII, L. O. (1957). *Ecology of Sea Colony Birds of the Barents Sea*. Israel Program for Scientific Translations. Jerusalem. 346 pp. (Translated from Russian, 1961.)

BENGSTON, S-A. (1984). Breeding ecology and extinction of the great auk (*Pinguinus impennis*): anecdotal evidence and conjectures. *Auk* 101: 1–12.

BERTRAM, G. C. L. & LACK, D. (1933). Notes on the birds of Bear Island. *Ibis* 3: 283–301.

BIGGAR, H. P. (1924). *The voyages of Jacques Cartier*. Publications of the Public Archives of Canada No. 11, 330 pp.

BIRKHEAD, T. R. (1976). Breeding biology and survival of guillemots *Uria aalge*. D. Phil. thesis, University of Oxford.

BIRKHEAD, T. R. (1977). The adaptive significance of the nestling period of guillemots *Uria aalge*. *Ibis*, 119. 544–549.

BIRKHEAD, T. R. (1978). Behavioural adaptations to high density nesting in the Common Guillemot *Uria aalge*. *Animal Behaviour*, 26: 321–331

BIRKHEAD, T. R. (1984). Distribution of the bridled form of the Common Guillemot *Uria aalge* in the North Atlantic. *Journal of Zoology, London*, 202: 165––76.

BIRKHEAD, T. R. & HARRIS, M. P. (1985). Ecological adaptations for breeding in the Atlantic Alcidae. In: *The Atlantic Alcidae* (Eds. D. N. Nettleship & T. R. Birkhead): pp 205–231. Academic Press, London.

BIRKHEAD, T. R., JOHNSON, S. D. & NETTLESHIP, D. N. (1985). Extra-pair matings and mate guarding in the common murre *Uria aalge*. Animal Behaviour, 33: 608–619.

BIRKHEAD, T. R., JOHNSON, S. D. & NETTLESHIP, D. N. (1986). Field observations of a possible hybrid murre *Uria aalge* × *Uria lomvia*. *Canadian Field Naturalist*, 100: 115–117.

BIRKHEAD, T. R. & LOCK, A. R. (1980). Changes in the proportion of bridled murres in Northern Labrador. *Condor* 82: 473–474.

BIRKHEAD, T. R. & MØLLER, A. P. (1992). *Sperm Competition in Birds: Evolutionary Causes and Consequences*. Academic Press, London.

BIRKHEAD, T. R. & NETTLESHIP, D. N. (1982). The adaptive significance of egg size and laying date in Thick-billed Murres *Uria lomvia*. *Ecology*, 63: 300–306.

BIRKHEAD, T. R. & NETTLESHIP, D. N. (1984). Egg size, composition and offspring quality in some Alcidae (Aves: Charadriiformes). *Journal of Zoology, London*, 202: 177–194.

BIRKHEAD, T. R. & NETTLESHIP, D. N. (1987a). Ecological relationships between common murres *Uria aalge* and thick-billed murres *U. lomvia* at the Gannet Islands, Labrador. I. Morphometrics and timing of breeding. *Canadian Journal of Zoology*, 65: 1621–1629.

BIRKHEAD, T. R. & NETTLESHIP, D. N. (1987b). Ecological relationships between common murres *Uria aalge* and thick-billed murres *U. lomvia* at the Gannet Islands, Labrador. II. Breeding success and site characteristics. *Canadian Journal of Zoology*, 65: 1630–1637.

BIRKHEAD, T. R. & NETTLESHIP, D. N. (1987c). Ecological relationships between common murres *Uria aalge* and thick-billed murres *U. lomvia* at the Gannet Islands, Labrador. III. Feeding ecology of the young. *Canadian Journal of Zoology*, 65: 1638–1649.

BIRKHEAD, T. R. & NETTLESHIP, D. N. (1988). Breeding performance of Black-legged kittiwakes, *Rissa tridactyla*, at a small, expanding colony in Labrador. *Canadian Field Naturalist*, 102: 20–24.

BIRKHEAD, T. R. & DEL NEVO, A. J. (1987). Egg formation and the pre- laying period of the Common Guillemot *Uria aalge*. *Journal of Zoology London*, 211: 83–88.

BIRKHEAD, T. R. & TAYLOR, A. M. (1977). Moult of the guillemot *Uria aalge*. *Ibis*, 119: 80–85.

BRADSTREET, M. S. W. & BROWN, R. G. B. (1985). Feeding Ecology of the Atlantic Alcidae. In: *The Atlantic Alcidae* (Eds. D. N. Nettleship & T. R. Birkhead): pp 263–318. Academic Press, London.

BRODY, H. (1975). *The People's Land: Eskimos and whites in the Eastern Arctic*. Penguin Books, Harmondsworth.

BRODY, H. (1987). *Living Arctic: Hunters of the Canadian North*. Faber & Faber, London.

BROWN, R. G. B. (1985). The Atlantic Alcidae at Sea. In: *The Atlantic Alcidae*, (Eds. D. N. Nettleship & T. R. Birkhead): pp 383–426. Academic Press, London.

BROWN, R. G. B., NETTLESHIP, D. N., GERMAIN, P., TULL, C. E. & DAVIS, T. (1975). *Atlas of Eastern Canadian Seabirds*. Canadian Wildlife Service, Ottawa, 220 pp.

BURGER, A. E. (1991). Maximum diving depths and underwater foraging in alcids and penguins. In: *Studies of High-Latitude Seabirds. 1. Behavioral, Energetic and Oceanographic aspects of Seabird Feeding Ecology*. (Eds. W. A. Montevecchi & A. J. Gaston): 9–15 Canadian Wildlife Service, Occasional Paper No. 68.

BURNESS, G. P. & MONTEVECCHI, W. A. (1992). Oceanographic-related variation in bone sizes of extinct great auks. *Polar Biology*, 11: 545–557.

BURKE, T. & BRUFORD, M. W. 1987. DNA fingerprinting in birds. Nature, London. 327: 149–152.
CARSCADDEN, J. (1984) Capelin (*Mallotus villosus*) in the Northwest Atlantic. In D. N. Nettleship, G. A. Sanger & P. F. Springer (eds), *Marine Birds: Their Feeding Ecology and Commercial Fisheries Relationships*, pp 170–183. Canadian Wildlife Service Special Publication, Ottawa.
CARTWRIGHT, F.D. (Ed.). (1826). *The Life and Correspondence of Major Cartwright*. London.
CARTWRIGHT, G. 1792. *Journal of Transactions and Events, during a residence of Nearly Sixteen Years on the Coast of Labrador; Containing Many Interesting Particulars, both of the Country and Its Inhabitants not Hitherto Known*. 3 Vols. Allin and Ridge, Newark, England.
CHARLESWORTH, M., FARRALL, L., STOKES, T. & TURNBULL, D. (1989). *Life Among the scientists: An Anthropological Study of an Australian Scientific Community*. Oxford University Press, Melbourne.
CHASTEL, C. E. (1988). Tick-borne virus infections of marine birds. *Advances in Disease Vector Research*, 5: 25–60.
CHIMMO, W. 1868. A visit to the North-East coast of Labrador, during the Autumn of 1867, by H.M.S. 'Gannet', Commander, W. Chimmo, R.N. *Royal Geographical Society Journal*, 38: 258–281.
CLUTTON-BROCK, T. H., GUINNESS, F. E. & ALBON, S. D. (1982). *Red Deer: Behavior and Ecology of Two Sexes*. Edinburgh University Press, Edinburgh.
COOK, F. A. (1911). *My attainment of the Pole*. Polar Publ. Co., New York.
CORDEAUX, J. (1872). *Birds of the Humber District*. Hull.
DAWKINS, R. (1976). *The Selfish Gene*. Oxford University Press, Oxford.
DAVIES, N. B. (1992). *Dunnock Behaviour and Social Evolution*. Oxford University Press, Oxford.
DENYS, N. (1672). *The Description and Natural History of the Coast of North America (Acadia)*. (Translated from French by W. F. Ganong, 1908). The Champlain Society, Toronto. 625 pp.
EAMES, H. (1973). *Winner Lose all: Dr Cook and the Theft of the North pole*. Little, Brown and Co. Boston.
ELLIOT, R. D. (1991). The management of the Newfoundland turr hunt. In: *Studies of High Latitude Seabirds. 2. Conservation Biology of Thick-billed Murres in the Northwest Atlantic*: 29–35. Canadian Wildlife Service Occasional Paper No. 69. Canadian Wildlife Service.
EVELEIGH, E. S. & THRELFALL, W. (1974). The biology of *Ixodes* (*Ceratixodes*) *uriae* White, 1852 in Newfoundland. *Acarologia*, t. XVI: 621-635.
FEILDEN, H. W. (1872). Birds of the Faeroe Islands. *Zoologist* 7 (Ser. 2): 3277–3294.
FISHER, J. & LOCKLEY, R. M. (1954). *Sea-birds*. New Naturalist, Collins, London.
FURNESS R. (1983). *The Birds of Foula*. Brathay, Ambleside.
GASTON, A. J. (1980). Populations, movements and wintering areas of Thick-billed murres (*Uria lomvia*) in eastern Canada. Canadian Wildlife Service Progress Note 110. 10 pp.
GASTON, A.J. (1984). How to distinguish first-year murres, *Uria* spp., from older birds in winter. *Canadian Field Naturalist* 98: 52–55.
GASTON, A. J. 1992. *The Ancient Murrelet: a Natural History in the Queen Charlotte Islands*. T. & A. D. Poyser, London.
GASTON, A. J., CAIRNS, D. K., ELLIOT, R. D. & NOBLE, D. G. (1985). *A Natural History of Digges Sound*. Canadian Wildlife Service Report No. 46. Ottawa.

GASTON, A. J., CHAPDELAINE, G., & NOBLE, D. G. (1983). The growth of thick-billed murre chicks at colonies in Hudson Strait: inter- and intra-colony variation. *Canadian Journal of Zoology*, 61: 2465–2475.

GASTON, A. J. & D. N. NETTLESHIP. (1981). *The Thick-billed Murres of Prince Leopold Island—a study of the breeding ecology of a colonial high arctic seabird.* Canadian Wildlife Service Monograph Series, No. 6. 350 pp.

GRAVES, J. A. & A. WHITEN (1980). Adoption of strange chicks by herring gulls *Larus argentatus* L. *Zeitschrift für Tierpsychologie*. 54: 267–278.

GREELY, A. W. (1886. *Three years of Arctic Service*. Scribner's Sons, New York.

GRIEVE, S. (1885). *The Great Auk or Garefowl* Alca impennis, *its History, Archaeology and Remains*. Thomas C. Jack. London, 142 pp.

GROVE, J. M. (1988). *The Little Ice Age*. Methuen, London.

HARRIS, M. P. (1970). Differences in the diet of British auks. *Ibis*, 112: 540–541.

HARRIS, M. P. & BIRKHEAD, T. R. (1985). Breeding ecology of the Atlantic Alcidae. In: *The Atlantic Alcidae* (Eds. D. N. Nettleship & T. R. Birkhead): Academic Press, London.

HARRIS, M. P. & WANLESS, S. (1984). The effects of disturbance on survival, age and weight of young guillemots *Uria aalge*. *Seabird*, 7: 42–46.

HARRIS, M. P., WEBB, A., & TASKER, M. L. (1991). Growth of young guillemots *Uria aalge* after leaving the colony. *Seabird*, 13: 40–44.

HARRISSE, H. (1900). *Decouverte et evolution cartographique de Terre-Neuve et des pays circonvoisins*. Revue de Geographie, Paris.

HATCHWELL, B. J. (1988). Intraspecific variation in extra-pair copulation and mate defence in common guillemots *Uria aalge*. *Behaviour*, 107: 157–185.

HATCHWELL, B. J. (1990). The effects of disturbance on the growth of young Common Guillemots *Uria aalge*. *Seabird*, 12: 35–39.

HATCHWELL, B. J. (1991). An experimental study of the effects of timing of breeding on the reproductive success of common guillemots (*Uria aalge*). *Journal of Animal Ecology*, 60: 721–736.

HEINRICH, B. (1990). *Ravens in winter*. Barrie & Jenkins, London.

HERSTEINSSON, P. (1984). The behavioural ecology of the arctic fox (*Alopex lagopus*) in Iceland. PhD. thesis University of Oxford. (Cited in: Kullberg, C. & Angerbjorn, A. (1992) Social behaviour and cooperative breeding in arctic foxes *Alopex lagopus* (L)., in a semi-natural environment. *Ethology*, 90: 321–355.)

HOBSON, K. A. & W. A. MONTEVECCHI (1991). Stable isotope determinations of trophic relationships of great auks. *Oecologia*, 87: 528–531.

HOMER, S. (1982). The quiet famine. *Equinox* 1: 42-57.

HUXLEY, J. (1939). Notes on the percentage of bridled guillemots. *British Birds*, 33: 174–183.

INGOLD, P. (1980). Anpassungen der Eier und des Brutverhaltens von Trottellummen an dasBruten auf Felssimsen. *Zeitschift für Tierpsychologie*, 53: 341–388.

JOHNSON, S. R. & WEST, G. C. (1975). Growth and development of heat regulation in nestlings, and metabolism of adult Common and Thick-billed Murres. *Ornis Scandinavica*, 6: 109–115.

JONES, T. R. (N.d., *c*, 1867). *Cassell's Book of birds*, Vol. III. Cassell, London.

JOURDAIN, F. C. R. (1922). The birds of Spitsbergen and Bear Island. *Ibis* 64: 159–179.

KOENIG, W. D. (1987). Reciprocal altruism in birds: a critical review. *Ethology Sociobiology*, 9: 73–84.

KREBS, J. R. & DAVIES, N. B. (1978) 2nd edition. *An Introduction to Behavioural Ecology*. Blackell, Oxford.

LANE, M. (1771). Directions for Navigating the Coast of Labrador between Spotted

Island and Sandwich Bay. *MS. Misc. Papers* Vol. 35: pp 1–6. Hydrographic Department, Taunton, Somerset.

LEIM, A. H. & SCOTT, W. B. (1966). *Fishes of the Atlantic coast of Canada*. Fisheries Research Board of Canada, Bulletin No. 155, 1–485.

LEY, W. (1935). The Great Auk. *Natural History*, 36: 351–357.

LIVEZEY, B. C. (1988). Morphometrics of flightlessness in the Alcidae. *Auk*, 105: 681–698.

LOCKLEY, R. M. (1930). *Dream Island*. H. F. & G. Witherby, London.

LOCKWOOD, W. B. (1984). *The Oxford book of British Bird Names*. Oxford University Press, Oxford.

LUCAS, F. A. (1890). *The Expedition to Funk Island, with Observations Upon the History and Anatomy of the Great Auk*. Report of the U. S. National Museum 1887–88: 493–529.

LYSAGHT, A.M. (1971). *Joseph Banks in Newfoundland and Labrador, 1766*. Fabee & Faber, London.

MCATEE, W. L. (2nd edition) 1959. *Folk Names of Canadian birds*. National Museum of Canada Bulletin 149.

McCLELLAND, D. 1962. On the dynamics of creative physical scientists. In: *The Ecology of Human intelligence* (Hudson, L., Ed): Penguin, Harmondsworth.

MACARTHUR, R. H. (1972). *Geographical Ecology*. Harper & Row, New York.

MACINTYRE, S. & SOOMAN, A. (1991). Non-paternity and prenatal genetic screening. *The Lancet*, 338: 869–871.

MACOUN, J. & MACOUN, J. M. (1909). Catalogue of Canadian Birds. Canada Department of Mines, Geological Survey Branch. No. 973, 1–761.

MARTIN, M. (1698). *A Late Voyage to St Kilda, the Remotest of All the Hebrides, or Western Isle of Scotland*. Gent, London. 159 pp.

MEARNS, B. & MEARNS, R. (1988). *Biographies for Birdwatchers: the Lives of Those Commemorated in West Palearctic Bird Names*. Academic Press, London.

MEDAWAR, P. (1982). *Pluto's Republic*. Oxford University Press, Oxford.

MELDGAARD, M. (1988). The great auk, *Pinguinus impennis* (L.) in Greenland. *Historical Biology* 1: 145–178.

MONAGHAN, P. (1992). Seabirds and sandeels: the conflict between exploitation and conservation in the northern North Sea. *Biodiversity & Conservation*. 1: 98–111.

MONTEVECCHI, W. A. & TUCK, L. M. (1987). *Newfoundland birds: Exploitation, Study, Conservation*. Nuttall Ornithological Club.

MORRIS, F. O. (1856). *A History of British Birds*. Groombridge, London.

MOSS, C. (1988). *Elephant Memories*. Elm Tree Books, London.

NANSEN, F. (1897). *Farthest North*. 2 Vols. Archibald Constable, Westminster.

NANSEN, F. (1911). *In Northern Mists*. 2 Vols. Ballantyne, London.

NELSON, J. B. (1978). *The Sulidae: Gannets and Boobies*. Oxford University Press, Aberdeen.

NELSON, T. H. (1907). *The Birds of Yorkshire*. 2 Vols. Brown & Sons, London.

NETTLESHIP, D. N. (1972). Breeding success of the common puffin (*Fratercula arctica* L). on different habitats at Great Island, Newfoundland. Ecology Monographs. 42: 239-268.

NETTLESHIP, D. N. (1980). A guide to major seabird colonies of eastern Canada: identity, distribution and abundance. Canadian Wildlife 'Studies on Northern Seabirds' Manuscript Report No. 97, 1–133.

NETTLESHIP, D. N. (1991). The diet of Atlantic Puffin chicks in Newfoundland before and after the initiation of an international capelin fishery, 1967–1984. *International Ornithology Congress*, XX: 2263–2271.

NETTLESHIP, D. N. in preparation. *Seabird Colonies of Northeastern North America*. Cornell University Press, Ithaca, New York.

NETTLESHIP, D. N. & EVANS, P. G. H. (1985). Distribution and status of the Atlantic Alcidae In: *The Atlantic Alcidae* (D. N. Nettleship, & T. R. Birkhead Eds): Academic Press, London.

NEWTON, A. (1861). Abstract of Mr. J. Wolley's researches in Iceland respecting the Gare-fowl or Great Auk (*Alca impennis*, L.). *Ibis*, 3: 374–399.

OLSON, S. L., SWIFT, C. C., & MOKHIBER, C. (1979). An attempt to determine the prey of the great auk (*Pinguinus impennis*). *Auk*, 96: 790–792.

OWEN, R. (1865). Description of the skeleton of the Great Auk, or Gare-Fowl (*Alca impennis*, L.). Transactions of Zoological Society of London. V: 317–335.

PARSONS, J. (1971). Cannibalism in Herring Gulls. *British Birds*, 64: 528–537.

PEARSON, T. H. (1968). The feeding ecology of seabird species breeding on the Farne Islands, Northumberland. *Journal of Animal Ecology*, 37: 521–552.

PIATT, J. F., NETTLESHIP, D. N. & THRELFALL, W. (1984). Net mortality of Common Murres *Uria aalge* and Atlantic Puffins *Fratercula arctica* in Newfoundland, 1951–1981. In: *Marine Birds: Their Feeding Ecology and Commercial Fisheries Relationships*, (D. N. Nettleship, G. A. Sanger and P. F. Springer Eds): 196–206. Canadian Wildlife Service Special Publication, Ottawa.

PIATT, J. F. & NETTLESHIP, D. N. (1985). Diving depths in four alcids. *Auk*, 102: 293–297.

PRINCE, P. A. & HARRIS, M. P. (1988). Food and feeding ecology of breeding Atlantic alcids and penguins. *Proceedings XIXth International Ornithology Congress*. Vol. 1: 1195–1204.

POTTER, S. & SARGENT, L. (1973). *Pedigree: Words from Nature*. New Naturalist, Collins, London.

RAY, J. (1678). *The Ornithology of Francis Willughby*. (Cited in Lockwood 1984.)

RIEDMAN, M. L. (1982). The evolution of alloparental care and adoption in mammals and birds. *Quarterly Revue of Biology*, 57: 405–435.

SALOMONSEN, F. (1945). Gejrfuglen et hundredaars minde. *Aarbog for Universitet Zioologiske Museum*, 1944–1945: 99–100.

SALOMONSEN, F. (1950) *Grønlands Fugle*. Munksgaard, Copenhagen. 604 pp.

SEEBOHM, H. (1896). *Coloured Figures of the Eggs of British Birds*. Pawson and Brailsford, Sheffield.

SEKORA, P. C., BYRD, G. V. & GIBSON, D. D. (1979). Breeding distribution and status of marine birds in the Aleutian Islands, Alaska. In: *Conservation of Marine Birds of Northern North America* (J. C. Bartonek, & D. N. Nettleship Eds): 33–39. U.S. Dept. of the Interior, Fish and Wildlife Service. Research Report 11. Washington, D.C.

SELOUS, E. (1901). *Birdwatching*. J. M. Dent. London.

SELOUS, E. (1905). *The Bird Watcher in the Shetlands*. J. M. Dent, London.

SERGEANT, D. E. 1951. Ecological relationships of the guillemots *Uria aalge* and *Uria lomvia*. Proceedings of the International Ornithology Congress. 10: 578–587.

SMITH, C. E. (1923). *From the Deep of the Sea: An Epic of the Arctic. The Diary of Charles Edward Smith*. Macmillan, New York.

SOUTHERN, H. N. (1962). Survey of bridled guillemots, 1959-1960. *Proceedings of the Zoological Society of London*. 138: 455–472.

STACEY, P. B. & KOENIG W. D. (1990). *Cooperative breeding in birds: long-term studies of ecology and behaviour*. Cambridge University Press, Cambridge.

STEELE, G. M., DAVIES, C. R., JONES, L. D., NUTTALL, P. A. & RIDEOUT, K. (1990). Life history of the seabird tick, *Ixodes* (*Ceratixodes*) *uriae*, at St. Abb's Head, Scotland. *Acarologia* 31: 125–130.

STEENSTRUP, J. (1868). Materiaux pour servir a l'histoire de l'Alca impennis (Linn.) et recherches sur les pays qu'il habitat. *Bulletin de la Societe Ornithologique Suisse* 2, Part 1, 5–70.
STONEHOUSE, B. (1960). The king penguin *Aptenodytes patagonica* of South Georgia. i. Breeding behaviour and development. *Falkland Islands Depend Surv Sci Report*, 6: 1–33.
STORER, R. W. (1960). Evolution of diving the birds. *Proceedings International Ornithological Congress*, 12: 694–707.
STORY, G. M., KIRWIN, W. J. & WIDDOWSON, J. D. A. (eds). (1990). *Dictionary of Newfoundland English* (2nd edn.). University of Toronto Press, Toronto.
STRAUCH, J. G. (1985). The phylogeny of the alcidae. *Auk* 102: 520–539.
SVERDRUP, O. (1904). *New Land*. Longman, Green & Co. London.
TANNER, V. (1947). Outlines of the Geography, life and customs of Newfoundland-Labrador. Cambridge University Press, Cambridge (originally published as Acta Geog. Helsinf. 8 No. 1).
TAUBER, P. (1875). On the fecundation of the egg in the common fowl. *Naturhistorisk Tidsskaift. Copenhagen*, 10: 63–106.
TAYLOR, C. E. & MCGUIRE, M. T. (1987). Reciprocal altruism: 15 years later. *Ethnology & Sociobiology*, 9: 67–72.
THOMAS, L. (1974). *The Lives of a cell*. Allen Lane, London.
TINBERGEN, N. (1958). *Curious Naturalists*. Country Life Ltd. London.
TOMPKINSON, P. M. L. & J. W. TOMPKINSON. (1966). Eggs of the great auk. *Bulletin of the British Museum (Natural History) Historical Series*, 3 (4): 97–128.
TOWNSEND, C. W. (Ed.). (1911). *Captain Cartwright and his Labrador Journal*. Estes & Co., Boston.
TSCHANZ, B. (1959). Zur brutbiologie der Trottellumme (*Uria aalge aalge* Pont.). *Behaviour*, 14: 1–108.
TSCHANZ, B. (1968). Trottellummen (*Uria aalge aalge* Pont.). *Zeitschrift für Tierpsychologie*, 4: 1–103.
TSCHANZ, B. (1979). Heifer-Beziehungen bei Trottellummen. *Zeitschrift für Tierpsychologie*, 49: 10–34.
TSCHANZ, B. & WEHRLIN, J. (1968). Krysning mellom lomvi, *Uria aalge* og polarlomvi, *Uria lomvia*, Pa Røst i Lofoten. *Fauna*, 21: 53–55.
TUCK, J. A. 1975. *The Archaeology of Saglek Bay, Labrador*. National Museums of Canada, Museum of Man, Mercury Series No. 32. Ottawa.
TUCK, J. A. (1976). *Ancient People of Port au Choix*. Newfoundland Social and Economic Studies No. 17. Memorial University of Newfoundland, St. John's. 262 pp.
TUCK, L. M. (1953). History and present populations of murre colonies in Newfoundland and Labrador. Canadian Wildlife Service Manuscript Report No. CWSC-665. 51 pp.
TUCK, L. M. (1961). The Murres. Canadian Wildlife Service, Ottawa.
VAROUJEAN, D. H., SANDERS, S. D., GRAYBILL, M. R. & SPEAR, L. (1979). Aspects of Common Murre breeding biology. *Pacific Seabird Group Bulletin*, 6: 28.
WADE, E. W. (1907). *The Birds of Bempton Cliffs: a Concise Description of the Different Species of Wild Birds that Frequent the Chalk Cliffs, with Full Details Respecting the Habits of the Guillemot*. Brown & Sons, London.
WANLESS, S. & HARRIS, M. P. (1985). Two cases of guillemots *Uria aalge* helping to rear neighbours' chicks on the Isle of May. *Seabird* 8: 5–8.
WHITBOURNE, R. (1622) *A Discourse and Discovery of Newfoundland*. Felix Kinston, London.

WHITELEY, W. H. (1964). The establishment of the Moravian Mission in Labrador and British Policy, 1763–83. *The Canadian Historical Review*, XLV: 29–50.

WILLIAMS, A. J. (1974). Site preferences and interspecific competition among guillemots (*Uria aalge* (L.)) and (*Uria lomvia* (L.)) on Bear Island. *Ornis Scandinavica*, 6: 117–124.

WILSON, E. O. (197)5. *Sociobiology: The New Synthesis*. Belknap Press, Harvard.

WILSON, E. O. (1987). *Biophilia*. Harvard University Press, Cambridge.

WHITEHEAD, H. & CARSCADDEN, J. E. (198)5. Predicting inshore whale abundance–whales and capelin off the Newfoundland coast. *Canadian Journal of Fisheries and Aquatic Science*, 42: 976 981.

WOOLER, R. D., BRADLEY, J. S. & CROXALL, J. P. (1992). Long-term population studies of seabirds. *Trends in Ecology and Evolution*, 7: 111–114.

YDENBERG, R. C. (1989). Growth-mortality trade-offs and the evolution of juvenile life histories in the Alcidae. *Ecology*, 70: 1494–1506.

ZIMAN, J. M. (1969). Information, communication and knowledge. *Nature*, 224: 318–324.

Index